DIFFERENTIAL GEOMETRY

William C. Graustein
Late Professor of Mathematics in Harvard University

DOVER PUBLICATIONS, INC.
Mineola, New York

Bibliographical Note

This Dover edition, first published in 1966 and republished in 2006, is an unabridged republication of the work originally published by The Macmillan Company, New York, in 1935.

Library of Congress Cataloging-in-Publication Data

Graustein, William C. (William Caspar), 1888–1941.
Differential geometry / William C. Graustein.
p. cm.
Originally published: New York : The Macmillan Company, 1935.
Includes index.
ISBN 0-486-45011-2 (pbk.)
1. Geometry, Differential. I. Title.

QA641.G7 2006
516.3'6—dc22

2005055596

Manufactured in the United States of America
Dover Publications, Inc., 31 East 2nd Street, Mineola, N.Y. 11501

PREFACE

This book furnishes an account, in terms of a vector notation, of the fundamentals of metric differential geometry of curves and surfaces in a Euclidean space of three dimensions. It contains essentially the material which the author gives in a half-year introductory course designed for the general student as well as for one intent on specialization in differential geometry.

The first nine chapters form a connected exposition of the theory, with simple illustrations. They treat, in addition to the fundamental properties of curves and surfaces, the mapping of surfaces and the absolute geometry on a surface. In the latter connection, the parallelism of Levi-Civita is introduced and the way is paved for the study of Riemannian geometry and its generalizations.

The tenth and last chapter has to do with the applications of the theory to certain important classes of surfaces.

No technical knowledge on the part of the reader is assumed beyond the fundamentals of the calculus, including partial differentiation, and the elements of solid analytic geometry. This does not mean that other tools are not at times employed. It is impossible to write on differential geometry without using the theory of differential equations, and certain theorems from the theory of functions are occasionally necessary if the treatment is to attain an aspect of rigor. But the precaution is taken always to state precisely the facts which are being assumed, so that the reader may not be hampered in his pursuit of the main line of thought.

In employing vector analysis in the development of any subject, there is a danger that the vector analysis will get out of hand and override the subject. Here, special care is taken to place the emphasis on the geometry, and to keep the vector analysis in its proper place as a tool. In fact, very little of what is generally understood as vector analysis is necessary. Only the algebra of vectors is used, and ten pages of the introduction suffice to give the reader all he needs to know about this topic.

In the algebra of vectors there are three combinations of vectors which are fundamental: the vector product of two vectors, the scalar product of two vectors, and the outer product, or determinant, of three vectors. These three products are connected by an important relation. If α, β, γ are three vectors, the determinant $|\alpha\,\beta\,\gamma|$ of their components is equal, for example, to the scalar product of the vector α with the vector product of the vectors β and γ.

From a logical point of view it would appear desirable that notations for the three products should take into account, first, the usual notation $|\alpha\,\beta\,\gamma|$ for a determinant, and, secondly, the basic relationship between the three products. The latter consideration would prescribe for the scalar product of two vectors a notation similar to that for a determinant. Actually, the notations of Grassmann conformed to these requirements. For the determinant of α, β, γ he wrote $[\alpha\,\beta\,\gamma]$, and for the scalar product of α and β he used $[\alpha|\beta]$.

In terms of these two symbols the fundamental relation between the three products would read $[\alpha\,\beta\,\gamma]=[\alpha|\beta\gamma]$, provided the vector product of β and γ is denoted by $\beta\gamma$. The transformation of $[\alpha\,\beta\,\gamma]$ into $[\alpha|\beta\gamma]$, or vice versa, would then be made by the insertion or deletion of a bar, and the fundamental relation would thus be rendered formally trivial.

The notation $\beta\gamma$ for a vector product, if it is not to lead to difficulties, needs some adornment. This should certainly not take the form of parentheses, inasmuch as parentheses have already been used to set off *scalar* products. Furthermore, it would seem fitting that it should emphasize the fact that the two vectors are actually being combined to form a single new vector. Such an adornment was conceived by Study in the shape of a "roof." His original symbol for the vector product was $\beta\,\hat{}\,\gamma$ and this has gradually evolved into $\widehat{\beta\,\gamma}$.

In the place of Grassmann's square brackets, Study introduced ordinary parentheses. He used, then, for the three products the symbols $(\alpha\,\beta\,\gamma)$, $(\alpha|\beta)$, and $\widehat{\alpha\,\beta}$ and accordingly wrote the fundamental relation between them in the form $(\alpha\,\beta\,\gamma) = (\alpha|\widehat{\beta\,\gamma})$.

I have employed Study's notation since my student days in Bonn, and use it in this book. In going into its motivation, my only purpose has been to outline the reasons for my conversion to it. There are other notations which are entirely adequate. If the reader is already familiar with one of them, I can assure him from personal experience that he will have no difficulty in translating the notation of Study into his own, and that his grasp of the algebra of vectors will undoubtedly be strengthened by the process.

It is with pleasure that I express my appreciation to The Macmillan Company for their efficient cooperation and courtesy.

W. C. G.

CAMBRIDGE, MASS.,
December, 1934.

CONTENTS

CHAPTER I. INTRODUCTION

CHAPTER II. SPACE CURVES

CHAPTER III. CURVES AND SURFACES ASSOCIATED WITH A SPACE CURVE

CHAPTER IV. FUNDAMENTALS OF THE THEORY OF SURFACES

CHAPTER V. CURVATURE. IMPORTANT SYSTEMS OF CURVES

CHAPTER VI. THE FUNDAMENTAL THEOREM

CHAPTER VII. GEODESIC CURVATURE. GEODESICS

CHAPTER VIII. MAPPING OF SURFACES

CHAPTER IX. THE ABSOLUTE GEOMETRY OF A SURFACE

CHAPTER X. SURFACES OF SPECIAL TYPE

DIFFERENTIAL GEOMETRY

CHAPTER I

INTRODUCTION

1. The nature of differential geometry. Differential geometry may be roughly described as the study of curves and surfaces of general type by means of the calculus. In contrast to it, there is *algebraic geometry*, which employs algebra as its principal tool and restricts itself to the consideration of a much narrower class of curves and surfaces. Thus, the theory of conic sections or quadric surfaces, with which the reader is familiar from analytic geometry, belongs to algebraic geometry, whereas that of the curvature of a general curve, or that of the tangent plane to a general surface, pertains to differential geometry.

A geometric configuration has two different kinds of properties, those which pertain to the configuration as a whole, and those which are definable for restricted portions of it. Thus, the order of a plane algebraic curve—the number of points in which it is cut by a straight line—, is a property of the curve in its entirety. On the other hand, the curvature of a curve at a point depends only on the shape of the curve in the neighborhood of the point.

Generally speaking, algebraic geometry is concerned with properties of the whole of a configuration, whereas differential geometry deals with properties of a restricted portion of it. Algebraic geometry is essentially a geometry of the whole or a *geometry in the large*, and differential geometry, *a geometry in the small.*

Euclidean geometry, either in the synthetic form of the preparatory school or the analytic form of the college, is algebraic geometry. So also is the ordinary projective geometry with which the reader is perhaps conversant. But these two geometries differ essentially in content. Euclidean geometry deals with properties of figures which are unchanged by rigid motions, for example, with distance, angle, and area. Projective geometry deals only with properties of figures which are unchanged by projections, such as the property that a point lie on a line, or that a number of lines be concurrent. The former is a quantitative, or metric, geometry, whereas the latter is concerned with properties of position and has nothing to do with measurement.

The distinction between metric and projective geometry is applicable, also, to differential geometry. Thus, there is a metric, or Euclidean, differential geometry and a projective differential geometry. In this book we shall be concerned only with metric differential geometry. In other words, we shall study, by means of the calculus, properties of curves and surfaces which are unchanged when the curves and surfaces are subjected to rigid motions.

We shall begin by recalling, in perhaps a somewhat novel form, a few facts from solid analytic geometry.

2. Directed line-segments. Vectors. As the basis of rectangular coordinates in space, we choose a system of coordinate axes which is right-handed, as shown in Fig. 1, and denote the coordinates of a point referred to this system by (x_1, x_2, x_3).

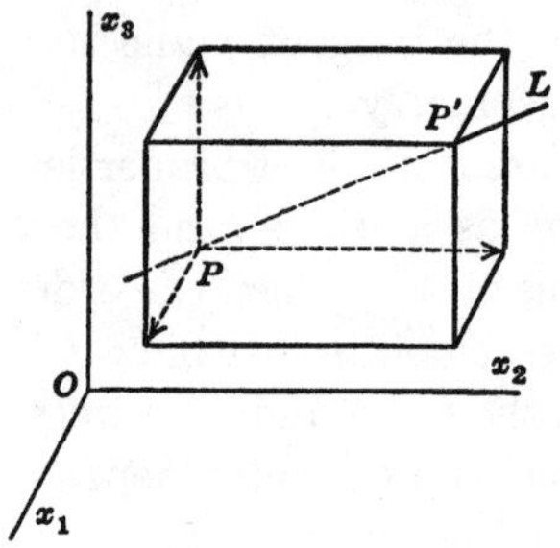

FIG. 1

Let $\overline{PP'}$ be the directed line-segment joining the point P with the coordinates (x_1, x_2, x_3) to the point P': (x'_1, x'_2, x'_3). If $\alpha_1, \alpha_2, \alpha_3$ are the measures of the projections of $\overline{PP'}$ on the axes of x_1, x_2, x_3, we know that

$$\alpha_1 = x'_1 - x_1, \qquad \alpha_2 = x'_2 - x_2, \qquad \alpha_3 = x'_3 - x_3. \tag{1}$$

In terms of $\alpha_1, \alpha_2, \alpha_3$, the length a of $\overline{PP'}$ is

$$a = \sqrt{\alpha_1^2 + \alpha_2^2 + \alpha_3^2}. \tag{2}$$

Finally, if A_1, A_2, A_3 are the angles (between 0 and π inclusive) which $\overline{PP'}$ makes with the positive axes of x_1, x_2, x_3, then

$$\alpha_1 = a \cos A_1, \qquad \alpha_2 = a \cos A_2, \qquad \alpha_3 = a \cos A_3. \tag{3}$$

By definition, $\cos A_1$, $\cos A_2$, $\cos A_3$ are the direction cosines of the line L on which $\overline{PP'}$ lies, when L is directed in the same sense as $\overline{PP'}$. Hence $\alpha_1, \alpha_2, \alpha_3$ are, themselves, direction components of the line L. We shall call them *the* direction components, or simply, the components, of the directed line-segment $\overline{PP'}$.

When $a = 1$, that is, when $\overline{PP'}$ is of unit length, $\alpha_1, \alpha_2, \alpha_3$ become the direction cosines of the directed line L, or, as we shall say, the direction cosines of the unit directed line-segment $\overline{PP'}$.

Vectors. It is evident geometrically that two directed line-segments which lie on the same line, or on parallel lines, and have the same sense and the same length, have the same components. Conversely, if $\alpha_1, \alpha_2, \alpha_3$ are three numbers, not all zero, there are infinitely many directed line-segments, in fact, one issuing from each point, which have $\alpha_1, \alpha_2, \alpha_3$ as their projections on the axes. Each two of these directed line-segments have the same direction, the same sense, and the same length.

It follows that, in order to fix the position in space of a directed line-segment $\overline{PP'}$, we must know, not only its components $\alpha_1, \alpha_2, \alpha_3$, but also the coordinates x_1, x_2, x_3 of its initial point P. If the components alone are known, the directed line-segment is free to move throughout space, provided merely that it keeps the same direction and sense. The directed line-segment is then called a *vector*. In other words, whereas a directed line-segment has precise position as well as direction, sense, and length, a vector has only direction, sense, and length. It takes all six quantities $x_1, x_2, x_3, \alpha_1, \alpha_2, \alpha_3$ to determine a directed line-segment, and only the three quantities $\alpha_1, \alpha_2, \alpha_3$ (not all zero) to determine a vector.

We shall call $\alpha_1, \alpha_2, \alpha_3$ the direction components, or simply, the components, of the vector; a, its length; and $\cos A_1, \cos A_2, \cos A_3$, its direction cosines. In short, we shall apply to vectors the terminology which we should naturally use for directed line-segments.

In order that every number triple $\alpha_1, \alpha_2, \alpha_3$ shall be the components of a vector, we introduce the *null* vector with the components 0, 0, 0. To distinguish other vectors from it, we shall call them *proper* vectors.

The vectors, proper and null, which we have thus far discussed are known as *free* vectors. In addition to them, we shall have occasion to use fixed, or *localized*, vectors. A localized vector is a vector with fixed initial point, that is, a directed line-segment. If the initial point is P, we shall speak of the vector as localized at P, or, more simply, as a vector at the point P.

3. Parallel and perpendicular vectors. Consider now a number of vectors $\alpha, \beta, \gamma, \cdots$, with the components $\alpha_1, \alpha_2, \alpha_3$, $\beta_1, \beta_2, \beta_3$, $\gamma_1, \gamma_2, \gamma_3, \cdots$, and think of them as being either all free vectors or all localized at the same point.

We think of two vectors as parallel when they have the same direction, provided they are proper vectors. When one is the null vector, the definition is inapplicable, since the null vector has no direction. To cover this case, we extend the definition by agreeing to regard the null vector as parallel to every vector.

If α is a proper vector and β is a vector parallel to α, the components of β are a multiple of those of α:

$$\beta_1 = k\alpha_1, \qquad \beta_2 = k\alpha_2, \qquad \beta_3 = k\alpha_3. \tag{4}$$

According as β is a proper vector with the same direction as α or the null vector, $k \neq 0$ or $k = 0$.

Conversely, equations (4) say that β is a vector parallel to α, the null vector if $k = 0$, and a proper vector if $k \neq 0$. Hence, we have proved the proposition:

THEOREM 1. *A necessary and sufficient condition that the vector β be parallel to the proper vector α is that there exist a constant k such that equations* (4) *are satisfied.*

In particular, if $k = -1$, β has the same length as α but the opposite sense; and, if $k = 1$, β is identical with α.

THEOREM 2. *The two vectors α and β are parallel if and only if*

$$\alpha_2\beta_3 - \alpha_3\beta_2 = 0, \quad \alpha_3\beta_1 - \alpha_1\beta_3 = 0, \quad \alpha_1\beta_2 - \alpha_2\beta_1 = 0. \tag{5}$$

In proving the theorem, we distinguish two cases, according as α is, or is not, the null vector.

If α is the null vector, equations (5) are obviously satisfied, on the one hand, and, on the other hand, α is, by definition, parallel to β.

If α is not the null vector, it suffices to show that equations (5) are equivalent to equations (4). That equations (5) follow from equations (4) is clear. Vice versa, equations (4) follow from equations (5). For, since α is not the null vector, at least one of its components is not zero. If $\alpha_3 \neq 0$, for example, the first two equations in (5) may be solved for β_2 and β_1, respectively. The resulting equations, $\beta_1 = (\beta_3/\alpha_3)\alpha_1$, $\beta_2 = (\beta_3/\alpha_3)\alpha_2$, together with $\beta_3 = (\beta_3/\alpha_3)\alpha_3$, are then precisely equations (4), when $k = \beta_3/\alpha_3$.

If we had not agreed to consider the null vector as parallel to every vector, the foregoing theorems would have been subject to exceptions which would later prove bothersome.

If α and β are both proper vectors, the angle θ between them, $0 \leqq \theta \leqq \pi$, is given by the familiar formula:

$$(6) \qquad \cos\theta = \frac{\alpha_1\beta_1 + \alpha_2\beta_2 + \alpha_3\beta_3}{\sqrt{\alpha_1^2 + \alpha_2^2 + \alpha_3^2}\sqrt{\beta_1^2 + \beta_2^2 + \beta_3^2}}.$$

The proper vectors α and β are, then, perpendicular if and only if $\alpha_1\beta_1 + \alpha_2\beta_2 + \alpha_3\beta_3 = 0$. Evidently, this equation is also satisfied if one of the vectors is the null vector. Accordingly, we agree to regard the null vector as perpendicular to every vector. We may, then, say:

THEOREM 3. *The vectors α and β are perpendicular if and only if*

$$(7) \qquad \alpha_1\beta_1 + \alpha_2\beta_2 + \alpha_3\beta_3 = 0.$$

We have found it expedient to regard the null vector as parallel, and also perpendicular, to every vector. We shall also think of it as parallel, and perpendicular, to every line and every plane. We shall not, however, think of it as making an angle with any other vector, line, or plane, inasmuch as formula (6) is meaningless if the length of either of the vectors in question is zero.

THEOREM 4. *If the vectors α and β are not parallel, then*

$$(8) \qquad \alpha_2\beta_3 - \alpha_3\beta_2, \quad \alpha_3\beta_1 - \alpha_1\beta_3, \quad \alpha_1\beta_2 - \alpha_2\beta_1$$

are the components of a proper vector which is perpendicular to each of them.

Since α and β are not parallel, the expressions (8) are, according to Theorem 2, not all zero, and hence are the components of a proper vector. That this vector is perpendicular to each of the vectors α and β may be proved by application of Theorem 3.

In this connection we note the identity of Lagrange:

$$(9) \quad (\alpha_2\beta_3 - \alpha_3\beta_2)^2 + (\alpha_3\beta_1 - \alpha_1\beta_3)^2 + (\alpha_1\beta_2 - \alpha_2\beta_1)^2$$
$$= (\alpha_1^2 + \alpha_2^2 + \alpha_3^2)(\beta_1^2 + \beta_2^2 + \beta_3^2) - (\alpha_1\beta_1 + \alpha_2\beta_2 + \alpha_3\beta_3)^2,$$

which is readily verified by expanding and comparing the two members.

Unit vectors. The vector α is a unit vector, that is, a vector of unit length, if and only if

$$\alpha_1^2 + \alpha_2^2 + \alpha_3^2 = 1.$$

The components $\alpha_1, \alpha_2, \alpha_3$ are, then, actually the direction cosines of the vector.

If α and β are unit vectors which are mutually perpendicular, $\alpha_1^2 + \alpha_2^2 + \alpha_3^2 = 1$, $\beta_1^2 + \beta_2^2 + \beta_3^2 = 1$, and $\alpha_1\beta_1 + \alpha_2\beta_2 + \alpha_3\beta_3 = 0$. Then the right-hand member of Lagrange's identity is equal to unity, and hence the vector with the components (8) is a unit vector. This result we may state as follows:

THEOREM 5. *If α and β are mutually perpendicular unit vectors, then*

$$(10) \quad \gamma_1 = \alpha_2\beta_3 - \alpha_3\beta_2, \quad \gamma_2 = \alpha_3\beta_1 - \alpha_1\beta_3, \quad \gamma_3 = \alpha_1\beta_2 - \alpha_2\beta_1$$

are the components of a unit vector, γ, which is perpendicular to each of them.

There is, of course, a second unit vector perpendicular to both α and β, namely, the vector with the components $-\gamma_1, -\gamma_2, -\gamma_3$. We shall call this the vector $-\gamma$.

The three vectors α, β, γ of Theorem 5 are mutually perpendicular unit vectors. We shall say that they have, in the order given, the same disposition as the coordinate axes if, after they have been localized at a point, there exists a rigid motion which carries α into OX_1, β into OX_2, and γ into OX_3, where OX_1, OX_2, OX_3 are the unit vectors at the origin of coordinates in the positive directions of the axes of x_1, x_2, x_3.

Consider, in addition to the triple of vectors α, β, γ, the triple $\alpha, \beta, -\gamma$. Only one of these two ordered triples can have the same disposition as the axes. This will be the triple α, β, γ or the triple $\alpha, \beta, -\gamma$ according as, after α and β have been varied continuously so as to become respectively OX_1 and OX_2, it is γ or $-\gamma$ which then becomes OX_3. But, when α and β have respectively the components 1, 0, 0 and 0, 1, 0, it is γ which has the components 0, 0, 1, as may readily be verified by means of formulas (10). Hence, it is the triple α, β, γ which has the same disposition as the axes.

This result we may state in the following way.

THEOREM 6. *If α, β, γ are three mutually perpendicular unit vectors which have, in the order given, the same disposition as the axes, then*

$$(10) \quad \gamma_1 = \alpha_2\beta_3 - \alpha_3\beta_2, \quad \gamma_2 = \alpha_3\beta_1 - \alpha_1\beta_3, \quad \gamma_3 = \alpha_1\beta_2 - \alpha_2\beta_1.$$

4. Algebra of number triples. The numbers with which we have been dealing, the coordinates of points and the components of vectors, appear in the form of ordered triples, and certain combinations of these triples enter frequently. It will be worth while to give to these combinations names, and to discuss their properties. In this connection we shall call a single number, to distinguish it from a triple of numbers, a scalar.

The two ordered triples of numbers a: (a_1, a_2, a_3) and b: (b_1, b_2, b_3) are identical if and only if $a_i = b_i$ $(i = 1, 2, 3)$. More often than not we shall write, instead of these three equations, the single symbolic equation $a = b$. Similarly, we shall use the symbolic equation $a = 0$ to stand for the three equations $a_i = 0$ $(i = 1, 2, 3)$ which say that the triple a is the triple $(0, 0, 0)$.

The scalar $a_1b_1 + a_2b_2 + a_3b_3$ is known as the *inner* or *scalar product* of the two triples a and b. It shall be denoted by $(a|b)$, —read "a into b":

$$(11) \qquad (a|b) = a_1b_1 + a_2b_2 + a_3b_3.$$

Evidently,

$$(b|a) = (a|b),$$

and

$$(a|a) = a_1^2 + a_2^2 + a_3^2.$$

The triple of two-rowed determinants $a_2b_3 - a_3b_2$, $a_3b_1 - a_1b_3$, $a_1b_2 - a_2b_1$ is called the *outer* or *vector product* of the triples a and b. We represent it by the symbol $\widehat{a\,b}$,—read "a b with a roof":

$$(12) \qquad \widehat{a\,b} : (a_2b_3 - a_3b_2,\ a_3b_1 - a_1b_3,\ a_1b_2 - a_2b_1).$$

Here,

$$\widehat{b\,a} = -\widehat{a\,b}, \qquad \widehat{a\,a} = 0.$$

The first of these equations stands for the three equations which state that the corresponding components of $\widehat{b\,a}$ and $\widehat{a\,b}$ are negatives of one another. Similarly, the second equation is symbolic for the fact that $\widehat{a\,a}$ is the triple (0, 0, 0).

The *determinant*

$$(13) \qquad (a\,b\,c) = \begin{vmatrix} a_1 & b_1 & c_1 \\ a_2 & b_2 & c_2 \\ a_3 & b_3 & c_3 \end{vmatrix}$$

is a combination of the three number triples a, b, c.

A fundamental relation. The development of the determinant $(a\,b\,c)$ according to the signed minors of the c's is the scalar product of the triple $\widehat{a\,b}$ with the triple c:

$$(a\,b\,c) = (\widehat{a\,b}\,|\,c).$$

For, the signed minors of c_1, c_2, c_3 are precisely the components $a_2b_3 - a_3b_2$, $a_3b_1 - a_1b_3$, $a_1b_2 - a_2b_1$ of the triple $\widehat{a\,b}$, and hence the development of $(a\,b\,c)$ according to them is

$$(a\,b\,c) = (a_2b_3 - a_3b_2)c_1 + (a_3b_1 - a_1b_3)c_2 + (a_1b_2 - a_2b_1)c_3 = (\widehat{a\,b}\,|\,c).$$

Similarly, $(a\,|\,\widehat{b\,c})$ is the development of $(a\,b\,c)$ by the signed minors of the a's: $(a\,|\,\widehat{b\,c}) = (a\,b\,c)$. Also, $(\widehat{c\,a}\,|\,b) = (c\,a\,b) = (a\,b\,c)$. Hence,

$$(14) \qquad (a\,b\,c) = (\widehat{a\,b}\,|\,c) = (\widehat{c\,a}\,|\,b) = (a\,|\,\widehat{b\,c}).$$

5. Applications to vectors. The results of § 3 may now be put in a strikingly simple form.

The vector α, with the components α_1, α_2, α_3, is a unit vector if and only if $(\alpha\,|\,\alpha) = 1$.

The angle θ, $0 \leqq \theta \leqq \pi$, between the two proper vectors α and β is given by

$$\cos\theta = \frac{(\alpha\,|\,\beta)}{\sqrt{(\alpha\,|\,\alpha)}\sqrt{(\beta\,|\,\beta)}},$$

or, if the vectors are unit vectors, by

$$\cos\theta = (\alpha\,|\,\beta).$$

The vectors α and β are perpendicular if and only if $(\alpha\,|\,\beta) = 0$. They are parallel if and only if $\widehat{\alpha\,\beta} = 0$; in particular, if α is a proper vector, β is parallel to α when and only when a scalar k

exists so that $\beta = k\,\alpha$, where $k\,\alpha$ is the vector whose components are k times those of α.

If α and β are not parallel, a proper vector perpendicular to them is $\widehat{\alpha\,\beta}$, and all the vectors perpendicular to them are given by $k\,\widehat{\alpha\,\beta}$, where k is an arbitrary scalar and $k\,\widehat{\alpha\,\beta}$ is the vector whose components are k times the components of $\widehat{\alpha\,\beta}$.

The foregoing statements cover Theorems 1–4 of § 3. Theorems 5 and 6 say that, if α and β are mutually perpendicular unit vectors, $\gamma = \widehat{\alpha\,\beta}$ is the unit vector perpendicular to each of them and so oriented that α, β, γ have the same disposition as the axes. But, then, these vectors, arranged in the order β, γ, α, also have the same disposition as the axes, and hence $\alpha = \widehat{\beta\,\gamma}$. Similarly, $\beta = \widehat{\gamma\,\alpha}$. Thus, the theorems in question may be extended, as follows:

THEOREM 1. *If α, β, γ are mutually perpendicular unit vectors which have, in the order given, the same disposition as the axes, then*

$$(15) \qquad \alpha = \widehat{\beta\,\gamma}, \qquad \beta = \widehat{\gamma\,\alpha}, \qquad \gamma = \widehat{\alpha\,\beta}.$$

Since $\widehat{\alpha\,\beta} = \gamma$, we have

$$(\alpha\,\beta\,\gamma) = (\widehat{\alpha\,\beta}\,|\,\gamma) = (\gamma\,|\,\gamma) = 1.$$

On the other hand, if α, β, γ had the disposition opposite to that of the axes, we should have $\widehat{\alpha\,\beta} = -\gamma$, and $(\alpha\,\beta\,\gamma)$ would have the value -1. In other words:

THEOREM 2. *A necessary and sufficient condition that the three mutually perpendicular unit vectors α, β, γ have, in the order given, the same disposition as the axes is that $(\alpha\,\beta\,\gamma) = 1$.*

We turn now to some propositions of a different type.

THEOREM 3. *The three vectors α, β, γ are parallel to a plane if and only if $(\alpha\,\beta\,\gamma) = 0$.*

If α, β, γ are parallel to a plane, $\widehat{\alpha\,\beta}$ is a vector perpendicular to the plane, and hence perpendicular to γ. Therefore $(\widehat{\alpha\,\beta}\,|\,\gamma) = 0$ and so $(\alpha\,\beta\,\gamma) = 0$.

Suppose, conversely, that $(\alpha\,\beta\,\gamma) = 0$ or $(\widehat{\alpha\,\beta}\,|\,\gamma) = 0$. If $\widehat{\alpha\,\beta} \neq 0$, $\widehat{\alpha\,\beta}$ is a proper vector perpendicular to γ, as well as to α and β, and hence α, β, γ are parallel to any plane perpendicular to $\widehat{\alpha\,\beta}$. If $\widehat{\alpha\,\beta} = 0$, α and β are parallel vectors and a plane parallel to one of them and γ is parallel to the other.

THEOREM 4. *The vectors* α, β, γ *are parallel to a plane if and only if scalars* k, l, m, *not all zero, exist so that*

$$k\alpha + l\beta + m\gamma = 0. \tag{16}$$

For, the three equations for which (16) is symbolic are homogeneous linear equations in k, l, m and have a solution for k, l, m, other than the solution 0, 0, 0, when and only when the determinant of their coefficients vanishes. But this is the determinant $(\alpha\,\beta\,\gamma)$, whose vanishing is the condition that the three vectors be parallel to a plane.

COROLLARY. *If the vectors* α, β, γ *are parallel to a plane and* α *and* β *are not parallel to one another, then scalars* A *and* B *exist so that*

$$\gamma = A\alpha + B\beta. \tag{17}$$

The vector γ is then said to be *a linear combination* of the vectors α and β.

It is evident that, if the scalar m is not zero, the symbolic relation (16) can be rewritten in the form (17) by dividing each of the equations for which it stands through by m and setting $-k/m = A$ and $-l/m = B$. But, if m were zero, the relation (16) would become $k\alpha = -l\beta$ and would say that the vectors α and β were parallel. Thus, the corollary is established.

THEOREM 5. *If* α, β, γ *are three vectors which are not parallel to a plane and* δ *is an arbitrarily chosen vector, three scalars,* A, B, C, *exist so that*

$$\delta = A\alpha + B\beta + C\gamma.$$

The theorem says that any vector can be written as a linear combination of three given vectors which are not parallel to a plane. We prove it by writing down the three equations

$$\begin{vmatrix} \alpha_i & \beta_i & \gamma_i & \delta_i \\ \alpha_1 & \beta_1 & \gamma_1 & \delta_1 \\ \alpha_2 & \beta_2 & \gamma_2 & \delta_2 \\ \alpha_3 & \beta_3 & \gamma_3 & \delta_3 \end{vmatrix} = 0, \qquad i = 1, 2, 3,$$

whose validity follows from the fact that the first row in each of the three determinants is identical with a subsequent row. When each determinant is expanded according to the minors of the elements in the first row, we have

$$(\beta\,\gamma\,\delta)\alpha_i - (\alpha\,\gamma\,\delta)\beta_i + (\alpha\,\beta\,\delta)\gamma_i - (\alpha\,\beta\,\gamma)\delta_i = 0, \quad i = 1, 2, 3.$$

Hence, we obtain the symbolic equation

$$(\alpha\,\beta\,\gamma)\delta = (\beta\,\gamma\,\delta)\alpha + (\gamma\,\alpha\,\delta)\beta + (\alpha\,\beta\,\delta)\gamma, \tag{18}$$

which, since $(\alpha\,\beta\,\gamma) \neq 0$, establishes the theorem.

6. The algebra of triples, continued. If k is a scalar and a is a triple, we shall mean by ka the triple ka_1, ka_2, ka_3. And, if a and b are triples, we shall mean by $a + b$ the triple $a_1 + b_1$, $a_2 + b_2$, $a_3 + b_3$.

It is readily verified that $(a|b)$, $\widehat{a\,b}$, and $(a\,b\,c)$ obey the following fundamental laws:

$$\begin{aligned} (ka|b) &= k(a|b), & (a + b|c) &= (a|c) + (b|c),\\ \widehat{ka\,b} &= k\,\widehat{a\,b}, & \widehat{a + b\;c} &= \widehat{a\,c} + \widehat{b\,c},\\ (ka\,b\,c) &= k(a\,b\,c), & (a + b\;c\,d) &= (a\,c\,d) + (b\,c\,d). \end{aligned}$$

As applications of these laws, we have, for example,

$$\begin{aligned} (ka + lb|c) &= (ka|c) + (lb|c) = k(a|c) + l(b|c),\\ (ka + lb|ka + lb) &= k^2(a|a) + 2kl(a|b) + l^2(b|b),\\ \widehat{ka + lb + mc\;d} &= k\,\widehat{a\,d} + l\,\widehat{b\,d} + m\,\widehat{c\,d}, \end{aligned}$$

where k, l, m are scalars and a, b, c, d are triples.

The generalized identity of Lagrange. The scalar product $(\widehat{a\,b}|\widehat{c\,d})$ of the vector products $\widehat{a\,b}$ and $\widehat{c\,d}$ can be expressed in terms of simple scalar products. We have, in fact, the identical relation:

$$(\widehat{a\,b}|\widehat{c\,d}) = (a|c)(b|d) - (a|d)(b|c), \tag{19}$$

which may be verified by expanding and comparing the two members.

When $c = a$ and $d = b$, the relation reduces to the identity of Lagrange, namely,

$$(\widehat{a\,b}|\widehat{a\,b}) = (a|a)(b|b) - (a|b)^2. \tag{20}$$

For simplicity, we shall usually suppress the "roofs" in $(\widehat{a\,b}|\widehat{c\,d})$ and $(\widehat{a\,b}|\widehat{a\,b})$ and write merely $(a\,b|c\,d)$ and $(a\,b|a\,b)$.

ω-identities. If the scalar product of a given triple a and an arbitrary triple ω vanishes for every choice of the triple ω:

$$(a|\omega) \equiv 0 \quad \text{or} \quad a_1\omega_1 + a_2\omega_2 + a_3\omega_3 \equiv 0, \tag{21}$$

the components of the triple a are all zero. For, since (21) holds for every triple ω, it holds for $\omega_1 = 1$, $\omega_2 = 0$, $\omega_3 = 0$, and hence $a_1 = 0$. Similarly, $a_2 = 0$, and $a_3 = 0$.

The first of the following theorems follows directly from these considerations and the second is a consequence of the first.

THEOREM 1. *A necessary and sufficient condition that a_1, a_2, a_3 be the triple* 0, 0, 0 *is that* $(a|\omega) \equiv 0$.

THEOREM 2. *The triples a and b are identical if and only if* $(a|\omega) \equiv (b|\omega)$.

To evaluate the vector product $\widehat{\widehat{a\,b}\,c}$ of the triple $\widehat{a\,b}$ with the triple c, we first simplify the scalar product $(\widehat{\widehat{a\,b}\,c}|\omega)$ of $\widehat{\widehat{a\,b}\,c}$ with the arbitrary triple ω. We have

$$(\widehat{\widehat{a\,b}\,c}|\omega) \equiv (\widehat{a\,b}\,c\,\omega) \equiv (\widehat{a\,b}|\widehat{c\,\omega}) \equiv (a|c)(b|\omega) - (b|c)(a|\omega),$$

or

$$(\widehat{\widehat{a\,b}\,c}|\omega) \equiv ((a|c)b - (b|c)a|\omega).$$

Hence, by Theorem 2,

$$\widehat{\widehat{a\,b}\,c} = (a|c)b - (b|c)a. \tag{22}$$

According to § 5, Theorem 4, this symbolic equation tells us that, if a, b, and $\widehat{a\,b}\,c$ are proper vectors at a point, the vector $\widehat{\widehat{a\,b}\,c}$ lies in the plane of the vectors a and b. How may this be proved geometrically?

It is evident from the deduction, in § 5, of the symbolic equation

$$(b\,c\,d)a - (a\,c\,d)b + (a\,b\,d)c - (a\,b\,c)d = 0, \tag{23}$$

that this equation is an identity in all four triples a, b, c, d. From it we obtain, by scalar multiplication by the triple e, the identity

$$(b\,c\,d)(a|e) - (a\,c\,d)(b|e) + (a\,b\,d)(c|e) - (a\,b\,c)(d|e) = 0 \tag{24}$$

in the five triples a, b, c, d, e.

EXERCISES

1. Verify the fundamental laws for $(a|b)$, $\widehat{a\,b}$, and $(a\,b\,c)$.

2. Verify the applications which are given of these laws.

3. Express $\widehat{\widehat{a\,b}\,c}$ as a linear combination of b and c.

4. Show that $(a\,b|c\,d) = (\widehat{a\,b}\,c\,d)$. Simplify $(a\,b\,\widehat{c\,d})$, and $(a\,\widehat{b\,c}\,d)$.

5. Prove that

$$(\widehat{a\,b}\,\widehat{c\,d}\,e) = (a\,c\,d)(b|e) - (b\,c\,d)(a|e) = (a\,b\,d)(c|e) - (a\,b\,c)(d|e).$$

6. Express the vector product of $\widehat{a\,b}$ and $\widehat{c\,d}$ as a linear combination of i) a and b; ii) c and d.

7. Find an expression for the length of the vector $\widehat{\widehat{\alpha\,\beta}\,\gamma}$.

8. Determine analytically and geometrically a proper vector which is parallel to the plane of the two nonparallel vectors α and β and is perpendicular to α.

9. Give a precise geometrical interpretation of the vector $\widehat{\widehat{\alpha\,\beta}\,\gamma}$, assuming that α and β are nonparallel vectors.

10. Let α, β, γ, δ be proper vectors at a point P not all lying in the same plane. Determine a proper vector at P lying on the line of intersection of the plane of α and β and the plane of γ and δ.

11. If x, y, z are triples whose components are functions of a parameter t, and x', for example, denotes the triple of derivatives with respect to t of the components of the triple x, show that

$$\frac{d}{dt}(x|y) = (x'|y) + (x|y'), \qquad \frac{d}{dt}\widehat{x\,y} = \widehat{x'\,y} + \widehat{x\,y'}.$$

Obtain a similar expression for the derivative of the determinant $(x\,y\,z)$.

7. Applications to solid analytic geometry. Let there be given at the point P: (x_1, x_2, x_3) three mutually perpendicular unit vectors, α, β, γ, which have the same disposition as the axes, and consider with reference to them the point Q: (y_1, y_2, y_3). Suppose that P is connected with Q by a broken line whose segments are parallel respectively to α, β, γ, and denote the algebraic lengths of these segments, referred to the positive senses of α, β, γ, by A, B, C; see Fig. 2. The projections on the axes of the broken line are then $A\alpha_i + B\beta_i + C\gamma_i$ $(i = 1, 2, 3)$. But these projections are equal respectively to the corresponding projections, $y_i - x_i$ $(i = 1, 2, 3)$, of the directed line-segment $\overline{PQ}$. Hence, we have, symbolically,

$$y = x + A\alpha + B\beta + C\gamma. \tag{25}$$

For example, if Q is the point on the line of the vector β which is at the algebraic distance r from P, where r is positive or negative according as $\overline{PQ}$ has the same sense as β or the opposite sense, then

$$y = x + r\beta. \tag{26}$$

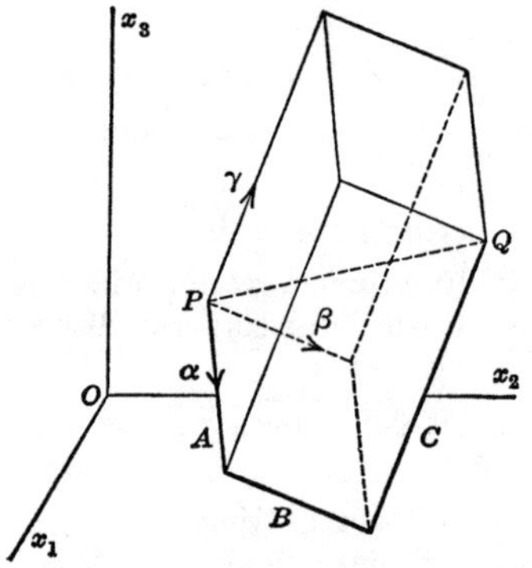

FIG. 2

We remark, incidentally, that, if we think of the lines of the vectors α, β, γ, directed in the same senses as the vectors, as constituting a new system of coordinate axes, then (A, B, C), or in the particular case $(0, r, 0)$, are the coordinates, referred to these axes, of Q.

The equation of the plane through the point r: (r_1, r_2, r_3) with a_1, a_2, a_3 as direction components of the normals, namely

$$a_1(x_1 - r_1) + a_2(x_2 - r_2) + a_3(x_3 - r_3) = 0,$$

can be written in the condensed form

$$(x - r | a) = 0.$$

The distance D from the point y: (y_1, y_2, y_3) to this plane is

$$D = \pm \frac{(y - r | a)}{\sqrt{a | a}}, \tag{27}$$

where, for the sake of simplicity of notation, we have written $a | a$ under the radical sign instead of $(a | a)$.

The straight line through the point r: (r_1, r_2, r_3) with the direction components a_1, a_2, a_3 is represented parametrically by the equations

$$x_1 = r_1 + a_1 t, \qquad x_2 = r_2 + a_2 t, \qquad x_3 = r_3 + a_3 t,$$

or by the single symbolic equation

$$x = r + at.$$

The square of the distance d from the point y: (y_1, y_2, y_3) to this line is

$$d^2 = \frac{(r - y\ a | r - y\ a)}{(a | a)}. \tag{28}$$

For, the square of the distance δ from the point (y_1, y_2, y_3) to an arbitrary point (x_1, x_2, x_3) on the line, namely

$$\delta^2 = (x - y | x - y) = (r - y + at | r - y + at),$$

or

$$\delta^2 = (r - y | r - y) + 2(r - y | a)t + (a | a)t^2, \tag{29}$$

is a minimum for the value of t which satisfies the equation,

$$2(r - y | a) + 2(a | a)t = 0,$$

obtained by equating to zero the derivative of δ^2 with respect to t. And, it is readily shown that, for this value of t, (29) reduces to (28).

CHAPTER II

SPACE CURVES

8. Parametric representation. There are two methods of representing a space curve analytically, either as the intersection of two surfaces or by means of a parametric representation. The latter method is better suited to our purposes.

The straight line through the point (r_1, r_2, r_3) with direction components a_1, a_2, a_3 has the parametric representation:

$$(1) \qquad x_1 = a_1t + r_1, \qquad x_2 = a_2t + r_2, \qquad x_3 = a_3t + r_3.$$

The circle in the (x_1, x_2)-plane whose center is at the origin and whose radius is a has the parametric equations

$$(2) \qquad x_1 = a \cos t, \qquad x_2 = a \sin t, \qquad x_3 = 0.$$

The equations

$$(3) \qquad x_1 = a \cos t, \qquad x_2 = a \sin t, \qquad x_3 = kt, \qquad k \neq 0,$$

represent parametrically the curve of the thread of a cylindrical screw, traced on the circular cylinder obtained by drawing the lines through the points of the circle (2) parallel to the x_3-axis. This curve is known as a circular helix. It furnishes a familiar example of a space curve which is *twisted,* that is, which does not lie in a plane.

Another example of a twisted curve is the twisted cubic:

$$(4) \qquad x_1 = at, \qquad x_2 = bt^2, \qquad x_3 = ct^3, \qquad abc \neq 0,$$

which takes its name from the fact that it is cut by a plane, in general, in three points, real or imaginary.

In the case of the straight line (1), there is just one point P of the line corresponding to a given value of t, and just one value of t corresponding to a given point P. We express these facts by saying that there is a one-to-one correspondence between the values of the parameter t and the points of the line.

The parameter t has, then, the nature of a coordinate: when t is given, a unique point P on the line is determined, and, when a point P on the line is given, t is uniquely determined. Accordingly, we shall call t a coordinate on the line.

The parameter t of the circular helix (3) or the twisted cubic (4) may also be thought of as a coordinate on the curve in question. In each of these cases, the parametric equations establish a one-to-one correspondence between the points of the curve and the values of t.

Equations (2) do not establish a one-to-one correspondence between the values of t and the points of the circle which they represent. There is a single point P of the circle corresponding to a given value of t. But there are infinitely many values of t which determine a given point P of the circle; if t_0 is one of them, the others are $t_0 \pm 2\pi$, $t_0 \pm 4\pi$, and so on. However, if the parameter t is restricted to the interval $0 \leqq t < 2\pi$, t is uniquely determined when P is given.

Consider, now, the three equations

$$(5) \qquad x_1 = x_1(t), \qquad x_2 = x_2(t), \qquad x_3 = x_3(t),$$

where $x_1(t)$, $x_2(t)$, $x_3(t)$ are three functions of the real variable t which we might, had we wished, have denoted by, say, $f(t)$, $\phi(t)$, $\psi(t)$. Assume that $x_1(t)$, $x_2(t)$, $x_3(t)$ are real, single-valued functions of t which are defined in a certain interval of values of t and are analytic in this interval, and rule out the case in which all three functions are constants. Then equations (5) represent, certainly, a space curve.

Since $x_1(t)$, $x_2(t)$, $x_3(t)$ are single-valued functions, a given value of t determines just one point of the curve. If necessary, we restrict the interval of values of t so that, vice versa, there is a unique value of t yielding a given point on the curve. Equations (5) then establish a one-to-one correspondence between the points of the curve and the values of t in the new interval, T, and t is a coordinate on the curve.

The demand that the functions in (5) be analytic in the interval T means that they are not only continuous in T, but have continuous derivatives of all orders. It means, further, that they may be represented in a neighborhood of an arbitrary but fixed point of T by power series, known as Taylor's series.

In the latter connection it is worth while to be more explicit. Let t be an arbitrarily chosen, but fixed, point in T; let x_1, x_2, x_3 be the values of the functions for this point t; x_1', x_2', x_3', the values of their first derivatives for the point t; x_1'', x_2'', x_3'', the values of their second derivatives for the point t; and so on. Then, if y_1, y_2, y_3 are the values of the functions for the point $t + \tau$, that is, if $y_1 = x_1(t + \tau)$, $y_2 = x_2(t + \tau)$, $y_3 = x_3(t + \tau)$, the three series,

$$
\begin{aligned}
y_1 &= x_1 + x_1'\tau + x_1''\frac{\tau^2}{2!} + x_1'''\frac{\tau^3}{3!} + \cdots, \\
y_2 &= x_2 + x_2'\tau + x_2''\frac{\tau^2}{2!} + x_2'''\frac{\tau^3}{3!} + \cdots, \qquad (6) \\
y_3 &= x_3 + x_3'\tau + x_3''\frac{\tau^2}{2!} + x_3'''\frac{\tau^3}{3!} + \cdots,
\end{aligned}
$$

converge and have the values y_1, y_2, y_3 for all values of τ which are numerically smaller than a properly chosen positive constant.

If P: (x_1, x_2, x_3) is the point of the curve (5) with the coordinate t, the point P': (y_1, y_2, y_3) with the coordinate $t + \tau$ is said to be a point in *the* neighborhood of P if τ lies in the common interval of convergence of the power series (6). *The* neighborhood of P may be quite extensive, and may, indeed, include the whole curve. Any part of it which contains P we shall call *a* neighborhood of P and any point in it we shall refer to as a point neighboring to P.

If equations (5) happen to be equations (1) subject to the restriction that t lie in the interval $0 \leqq t \leqq 1$, they represent only a line-segment. Thus, parametric equations may define only a portion of a curve. We agree, however, always to speak of a curve and not of a curve-segment.

9. Regular curves and regular parameters. A space curve which admits a representation by equations of the form (5), where the functions are subject to the conditions prescribed, is known as an *analytic space curve.*

An analytic space curve for which there exists a representation by functions $x_1(t)$, $x_2(t)$, $x_3(t)$ whose first derivatives, x_1', x_2', x_3', never vanish simultaneously is called a *regular analytic space curve,* and the parameter t in terms of which it is thus represented is known as a *regular parameter.*

If the first derivatives of the functions in (5) all vanish for a certain value of t, we cannot conclude, without further investigation, that the analytic curve represented by equations (5) is not regular. It may be that it is the parameter t, and not the curve, which fails to be regular. This is the fact in the case of the curve $x_1 = t^3$, $x_2 = t^6$, $x_3 = t^9$, for this curve is equally well represented by the regular parametric equations, $x_1 = t$, $x_2 = t^2$, $x_3 = t^3$. On the other hand, the curve $x_1 = t^2$, $x_2 = t^3$, $x_3 = t^4$, though analytic, is not regular; it is impossible to find for it a regular parametric representation.

We shall assume always that equations (5) define a regular analytic curve expressed parametrically in terms of a regular parameter. Instead of these equations we shall ordinarily write the single symbolic equation $x = x(t)$. Similarly, we shall use, in place of the triple of derivatives x_1', x_2', x_3', the symbolic derivative x', and, in place of equations (6), the symbolic equation

$$(7) \qquad y = x + x'\tau + x''\frac{\tau^2}{2!} + x'''\frac{\tau^3}{3!} + \cdots .$$

Change of parameter. Consider the relation $t = f(u)$, where $f(u)$ is a real, single-valued, analytic function in an interval, U, of values of the real variable u, and assume that the values of t obtained when u takes on all the values in U lie in and exhaust the interval T defined on p. 17. Then, the equations obtained from (5) by substituting $f(u)$ for t constitute a new representation of the curve.

Since the new parametric equations, $x = \bar{x}(u)$, result from $x = x(t)$ by setting $t = f(u)$, we have $\bar{x}(u) = x(t)$, where $t = f(u)$. It follows that $\bar{x}' = f'(u)x'$, where $\bar{x}' = d\bar{x}/du$ and $x' = dx/dt$. Consequently, since $x' \neq 0$ by hypothesis, a necessary and sufficient condition that $\bar{x}' \neq 0$, that is, that $\bar{x}_1'$, $\bar{x}_2'$, $\bar{x}_3'$ never vanish simultaneously, is that $f'(u)$ does not vanish. Thus we have proved the proposition:

THEOREM 1. *If t is a regular parameter, then u is a regular parameter if and only if dt/du never vanishes.*

From analysis we know that the demand that dt/du never vanish in the interval U means that the function $u = \phi(t)$, which is the inverse of the given function $t = f(u)$, is single-valued and analytic in the interval T.

Since the given function and its inverse are single-valued, there is established a one-to-one correspondence between the values of t in T and the values of u in U. Hence, there is a one-to-one correspondence between the values of u in U and the points of the curve, and u is a coordinate on the curve.

A characterization of regular curves. At a point P of the curve $x = x(t)$, at least one of the derivatives, x_1', x_2', x_3', is not zero. If x_1', for example, is not zero at P, it follows, since $x_1'(t)$ is a continuous function, that there exists a certain neighborhood of P at every point of which x_1' is not zero. Consequently, by Theorem 1, x_1 is a regular parameter for this piece of the curve: there exists for the piece a representation of the form $x_1 = x_1$, $x_2 = f(x_1)$, $x_3 = \phi(x_1)$, where $f(x_1)$, $\phi(x_1)$ are single-valued, analytic functions of x_1.

The converse is also true. Hence, an analytic curve is regular if and only if, at each point of the curve, at least one of the coordinates x_1, x_2, x_3 is a regular parameter for a certain neighborhood of the point.

EXERCISES

1. Show that the curve $x_1 = t$, $x_2 = t^2$, $x_3 = t^4$ lies on each of the parabolic cylinders $x_2 = x_1^2$, $x_3 = x_2^2$, and on the quartic cylinder $x_3 = x_1^4$. Determine the complete intersection of each two of these surfaces.

2. Show that the curve

$$x_1 = a \sin^2 t, \quad x_2 = a \sin t \cos t, \quad x_3 = a \cos t$$

lies on a sphere, a circular cylinder, and a parabolic cylinder, and is the complete intersection of any two of these surfaces.

3. Write a parametric representation of the curve in which the circular cylinder $x_1^2 + x_2^2 = a^2$ meets the hyperbolic paraboloid $2\,x_1x_2 = ax_3$.

10. Length of arc. The direction along the curve

$$x = x(t) \tag{8}$$

in which the curve is traced when the parameter t increases is called the positive direction along the curve, and the opposite direction, the negative direction.

Let P_0, with the coordinate t_0, be a fixed point of the curve and let P, with the coordinate t, be an arbitrary point so chosen that the direction from P_0 to P along the curve is the positive direction. Mark on the arc P_0P the succession of points P_1, P_2, $\cdots$, P_{n-1} and draw the chords P_0P_1, P_1P_2, $\cdots$, $P_{n-1}P_n$, where

P_n is P. Then the limit of the length of the broken line consisting of these chords, when n becomes infinite so that the length of the longest chord approaches zero, exists and is defined as the length of arc, s, of the curve from P_0 to P:

$$s = \lim_{n\to\infty} \sum_{i=1}^{n} P_{i-1}P_i. \tag{9}$$

As a matter of fact, it can be shown, by the Integral Calculus, that

$$s = \int_{t_0}^{t} \sqrt{(x_1')^2 + (x_2')^2 + (x_3')^2}\,dt = \int_{t_0}^{t} \sqrt{x'\,|\,x'}\,dt. \tag{10}$$

If we had taken P so that the direction along the curve from P_0 to P was the negative direction, we should have inserted a minus sign on the right-hand side of (9). The length of arc is, then, positive or negative according as it is measured in the positive or negative direction along the curve.

Formula (10) gives the correct result in both cases, for the integral is positive or negative according as $t > t_0$ or $t < t_0$.

From (10) we have

$$ds = \sqrt{x'\,|\,x'}\,dt. \tag{11}$$

Since $dx = x'dt$, $(dx\,|\,dx) = (x'\,|\,x')dt^2$, and hence

$$ds^2 = (dx\,|\,dx) = dx_1^2 + dx_2^2 + dx_3^2. \tag{12}$$

Equation (10) defines s as a function of t: $s = f(t)$. This function is analytic, and, since $(x'\,|\,x')$ is never zero, its first derivative, ds/dt, never vanishes. The inverse function $t = \phi(s)$ is, then, single-valued and analytic and its first derivative is never zero.

Hence, we may set $t = \phi(s)$ for t in (8), to obtain a parametric representation of the given curve in terms of the arc s. In other words, s is a regular parameter for the curve.

It is therefore conceivable that the parameter t is itself the arc of the curve, measured from a suitable point. This is the case if and only if $(x'\,|\,x')$ is constant and equal to unity.

THEOREM 1. *A necessary and sufficient condition that t be the arc of the curve is that*

$$(x'\,|\,x') \equiv 1. \tag{13}$$

We are now in a position to prove the important proposition:

THEOREM 2. *If P' is a point on the curve neighboring to the point P, the ratio of the chord PP' to the arc PP' approaches unity when P' approaches P along the curve.*

Let the curve be represented in terms of the arc as parameter (or coordinate), and let P and P' be the points with the coordinates s and $s + \sigma$, respectively. Then the (directed) arc PP' is σ.

Employing Taylor's series, we write symbolically, for the coordinates (y_1, y_2, y_3) of P',

$$y = x + x'\sigma + x''\frac{\sigma^2}{2!} + x'''\frac{\sigma^3}{3!} + \cdots, \tag{14}$$

where $x, x', x'', \cdots$ are symbolic for the values of the functions $x(s)$ and their derivatives at the point P.

For the chord length PP' we have:

$$\overline{PP'}^2 = (y - x \mid y - x) = \left(x'\sigma + x''\frac{\sigma^2}{2!} + \cdots \middle| x'\sigma + x''\frac{\sigma^2}{2!} + \cdots\right),$$

or

$$\overline{PP'}^2 = (x' \mid x')\sigma^2 + (x' \mid x'')\sigma^3 + \cdots. \tag{15}$$

Hence

$$\left(\frac{PP'}{\sigma}\right)^2 = (x' \mid x') + (x' \mid x'')\sigma + \cdots,$$

and

$$\lim_{\sigma \to 0} \left(\frac{PP'}{\sigma}\right)^2 = (x' \mid x').$$

But $(x' \mid x') = 1$, by Theorem 1, and the proof is complete.

In deducing (15), we squared each of the three power series represented by (14) and added the resulting power series term by term. Though the justification of these operations is of prime importance to us, it belongs in a course in analysis. Accordingly, we content ourselves, here and later, with the knowledge that we can safely do anything with power series, within their domains of convergence, that we can do with polynomials.

Example 1. In the case of the circular helix (3), the triple x' has the components $-a \sin t, a \cos t, k$. Hence, $(x' \mid x') = a^2 + k^2$, and the arc, s, measured from the point for which $t = 0$, is

$$s = \int_0^t \sqrt{a^2 + k^2}\,dt = \sqrt{a^2 + k^2}\,t.$$

Hence, $t = s/\sqrt{a^2 + k^2}$, and parametric equations of the curve, in terms of the arc as parameter, are

$$x_1 = a \cos (s/\sqrt{a^2+k^2}), \quad x_2 = a \sin (s/\sqrt{a^2+k^2}), \quad x_3 = (ks)/\sqrt{a^2+k^2}.$$

Example 2. For the twisted cubic $x_1 = 6\,t$, $x_2 = 3\,t^2$, $x_3 = t^3$, it may be shown that $(x' \,|\, x') = (3\,t^2 + 6)^2$, and hence that the arc s, measured from the point $t = 0$, is: $s = t^3 + 6\,t$. It follows that

$$t = \sqrt[3]{(s/2) + \sqrt{(s/2)^2 + 8}} + \sqrt[3]{(s/2) - \sqrt{(s/2)^2 + 8}},$$

and substitution of this value of t in the equations of the curve would yield a representation of the curve in terms of the arc as parameter.

Example 3. For the general twisted cubic (4), we find that

$$s = \int_0^t \sqrt{a^2 + 4\,b^2t^2 + 9\,c^2t^4}\,dt.$$

This integral is, in general, an elliptic integral and cannot be evaluated in terms of the elementary functions.

The last two examples teach us that the introduction of the arc as parameter, though always possible and theoretically of great value, is in the case of particular curves seldom practical or of advantage.

11. The derived vectors. In the future, in order to avoid ambiguity, we shall distinguish between derivatives with respect to t and derivatives with respect to s by attaching a subscript s to the latter derivatives. Thus, x'_s shall mean the triple of first derivatives of the functions x with respect to s, whereas x'' shall denote the triple of second derivatives of the functions x with respect to t.

The triple $x'(t)$ defines a vector, the vector with the components $x'_1(t)$, $x'_2(t)$, $x'_3(t)$. This vector we think of as localized at the point P of the curve whose coordinate is t. Similarly, we may interpret x'', x''', $\cdots$ and x'_s, x''_s, x'''_s, $\cdots$ as vectors at the point P.

Since we have assumed that x'_1, x'_2, x'_3 do not vanish simultaneously for any value of t, the vector x' is a proper vector at every point of the curve.

It is evident, from Theorem 1 of § 10, that

$$(x'_s \,|\, x'_s) \equiv 1. \tag{16}$$

Differentiating this identity with respect to s, we obtain the new identity:

$$(x'_s \,|\, x''_s) \equiv 0. \tag{17}$$

By means of (16) and (17), it is readily verified that

$$(x'_s\, x''_s \,|\, x'_s\, x''_s) \equiv (x''_s \,|\, x''_s). \tag{18}$$

Equations (16) and (17) tell us that the vector x'_s is of unit length, and that the vector x''_s is perpendicular to the vector x'_s.

12. Tangent line. The tangent line at a point P of a curve is the limiting position of the secant joining P to a neighboring point P', when P' approaches P along the curve. As the positive direction on the tangent we take that direction which coincides at P with the positive direction on the curve. We then direct the secant PP' so that the directed tangent will be the limit of the directed secant. This means, evidently, that the secant is to be directed from P to P' or from P' to P according as the direction along the curve from P to P' is the positive or the negative direction.

The points P and P', corresponding respectively to the values s and $s + \Delta s$ of the arc, have the space coordinates (x_1, x_2, x_3) and $(x_1 + \Delta x_1, x_2 + \Delta x_2, x_3 + \Delta x_3)$, where Δx_1, Δx_2, Δx_3 are the increments of x_1, x_2, x_3 which correspond to the increment Δs of s. The direction cosines of the directed secant PP' are, then,

$$\pm \frac{\Delta x_1}{PP'}, \qquad \pm \frac{\Delta x_2}{PP'}, \qquad \pm \frac{\Delta x_3}{PP'},$$

where the upper, or the lower, signs are to be taken according as $\Delta s > 0$ or $\Delta s < 0$. But

$$\lim_{\Delta s \to 0} \left(\pm \frac{\Delta x_i}{PP'} \right) = \lim_{\Delta s \to 0} \frac{\Delta x_i}{\Delta s} \lim_{\Delta s \to 0} \left(\pm \frac{\Delta s}{PP'} \right) = \frac{dx_i}{ds}, \quad i = 1, 2, 3.$$

Hence the direction cosines of the directed tangent, which we shall denote by α_1, α_2, α_3, are

$$\alpha_1 = \frac{dx_1}{ds}, \qquad \alpha_2 = \frac{dx_2}{ds}, \qquad \alpha_3 = \frac{dx_3}{ds},$$

or, symbolically,

$$\alpha = x'_s. \tag{19}$$

When the curve is defined parametrically in terms of an arbitrary parameter t, $ds = \sqrt{x'|x'}\,dt$ and therefore

$$x'_s = \frac{dx}{dt}\frac{dt}{ds} = \frac{x'}{\sqrt{x'|x'}}.$$

Thus, the expression for α in terms of derivatives with respect to the parameter t is

$$\alpha = \frac{x'}{\sqrt{x'|x'}}. \tag{20}$$

The vector α at the point P is a unit vector which lies on, and has the same sense as, the directed tangent. We shall call it *the tangent vector* at P.

The assumption that x' is never zero implied in § 10 the possibility of introducing the arc as parameter throughout the entire extent of the curve. It implies here the validity, at every point of the curve, of the foregoing theory of the tangent line.

Example. For the circular helix (3), x' has the components $-a \sin t$, $a \cos t$, k and $(x'|x') = a^2 + k^2$. Hence, the direction cosines of the tangent at an arbitrary point P, as given by (20), are readily written down. In particular, α_3 has the value $k/\sqrt{a^2 + k^2}$ and so is constant. But α_3 is the cosine of the angle which the directed tangent at P makes with the ruling of the cylinder which passes through P. Hence, the helix cuts the rulings of the cylinder under a constant angle.

It is convenient to introduce this angle, θ, in place of the constant k. From $\cos\theta = k/\sqrt{a^2 + k^2}$, it follows that $k = a \cot\theta$. Hence, the parametric equations of the curve become

$$x_1 = a \cos t, \quad x_2 = a \sin t, \quad x_3 = a \cot\theta\, t,$$

and α_1, α_2, α_3 have the values $-\sin\theta \sin t$, $\sin\theta \cos t$, $\cos\theta$.

EXERCISE

Find the points in which the twisted cubic of § 10, Example 2, meets the plane

$$x_1 + 4\,x_2 - 6\,x_3 - 12 = 0$$

and the angle at which it cuts the plane at one of the points.

13. Contact of the tangent line with the curve. According to (20), the tangent line at P: (x_1, x_2, x_3) has the direction components x_1', x_2', x_3'. Hence the distance to it from a point P': (y_1, y_2, y_3) on the curve neighboring to P is given by the formula (§ 7)

$$d^2 = \frac{(\widehat{x'\, y - x}\,|\,\widehat{x'\, y - x})}{(x'\,|\,x')}\,.$$

If t and $t + \tau$ are the coordinates of P and P',

$$y = x + x'\tau + x''\frac{\tau^2}{2!} + x'''\frac{\tau^3}{3!} + \cdots.$$

Then

$$\widehat{x'\, y - x} = \widehat{x'\,x''}\frac{\tau^2}{2!} + \widehat{x'\,x'''}\frac{\tau^3}{3!} + \cdots,$$

and

$$(21) \qquad (x'\,|\,x')d^2 = (\widehat{x'\,x''}\,|\,\widehat{x'\,x''})\frac{\tau^4}{4} + \cdots.$$

If $(\widehat{x'x''}\,|\,\widehat{x'x''}) \neq 0$ at P, that is, if $\widehat{x'\,x''} \neq 0$ at P, d^2 is an infinitesimal of the fourth order with respect to τ. But τ and the arc PP', or Δs, are infinitesimals of the same order, since $\lim\,(\Delta s/\tau) = ds/dt \neq 0$. Thus, d is an infinitesimal of the second order with respect to the arc PP'. We say, then, that the tangent line at P has contact of the first order with the curve.

It follows from (21) that the tangent line at P has contact of at least the second order with the curve, that is, that d is an infinitesimal of at least the third order with respect to the arc PP', when and only when $\widehat{x'\,x''} = 0$ at P.

If the tangent line at every point has contact of at least the second order with the curve, the curve is a straight line. For, if $\widehat{x'\,x''} \equiv 0$, then at every point of the curve the vector x'' is parallel to the vector x'. Hence, since x' is always a proper vector, a scalar function $f(t)$ exists so that

$$x'' = f(t)x',$$

or

$$\frac{dx'}{x'} = f(t)dt.$$

Integrating, we have

$$\log x' = \int f(t)dt + \log a \quad \text{or} \quad x' = ae^{\int f(t)dt},$$

where a is a triple of constants. Hence

$$x = a\left(\int e^{\int f(t)dt} dt\right) + r,$$

where r is a second triple of constants. But this symbolic equation, when the scalar function in the parenthesis is replaced by the new parameter u, becomes

$$x = au + r, \tag{22}$$

and therefore represents a straight line.

Conversely, if $x = x(t)$ represents a straight line, then $\widehat{x' x''} \equiv 0$. For, every parametric representation of the arbitrary line (22) is of the form $x = a\phi(t) + r$, where $\phi(t)$ is a scalar function of t. Now $x' = \phi'(t)\, a$, $x'' = \phi''(t)\, a$, and $\widehat{x' x''} = \phi'\phi''\widehat{a\, a} \equiv 0$.

We have thus proved the important proposition:

THEOREM 1. *A necessary and sufficient condition that the curve* $x = x(t)$ *be a straight line is that* $\widehat{x' x''} \equiv 0$.

According as $\widehat{x' x''} \not\equiv 0$ or $= 0$ at a point of a curve, not a straight line, we shall call the point a *regular point* or a *singular point*. In the next few paragraphs, we shall restrict ourselves primarily to regular points.

Closely connected with Theorem 1 is the following theorem which we shall find useful later.

THEOREM 2. *A necessary and sufficient condition that all the vectors of the one-parameter family of proper vectors defined by the symbolic equation* $a = a(t)$ *have the same direction is that* $\widehat{a\, a'} \equiv 0$.

If the vectors all have the same direction, the vectors $a = a(t)$ and $a + \Delta a = a(t + \Delta t)$ are always parallel: $\widehat{a\; \overline{a + \Delta a}} \equiv 0$. Hence, $\widehat{a\, \Delta a} \equiv 0$, or $\widehat{a\, \Delta a}/\Delta t \equiv 0$, and therefore $\widehat{a\, a'} \equiv 0$.

Conversely, if $\widehat{a\, a'} \equiv 0$, a scalar function $f(t)$ exists so that $a' = f(t)a$. Hence, $a = \phi(t)c$, where c is a triple of constants and $\log \phi(t) = \int f(t)dt$. Thus, the vectors $a(t)$ all have the direction of the vector c.

When interpreted in the light of Theorem 2, Theorem 1 says that the curve $x = x(t)$ is a straight line if and only if the tangent vector $x'(t)$ always has the same direction.

EXERCISE

The tangent line at P has contact of order m with the curve if and only if at P the vectors $x', x'', \cdots, x^{[m]}$ are collinear, but $x^{[m+1]}$ does not lie on their line.

14. Osculating plane. We seek, if it exists, the plane through the regular point P: (x_1, x_2, x_3) of a curve, not a straight line, which has contact of higher order with the curve than any other plane through the point. To this end, we discuss the infinitesimal distance, D, from a neighboring point P': (y_1, y_2, y_3) of the curve to an arbitrarily chosen, but fixed, plane through P. If a is a vector normal to the plane, then (§ 7)

$$D = \pm \frac{(y - x \mid a)}{\sqrt{a \mid a}}.$$

Since

$$y = x + x'\tau + x''\frac{\tau^2}{2!} + x'''\frac{\tau^3}{3!} + \cdots,$$

we have

$$\pm \sqrt{a \mid a}\, D = (x' \mid a)\tau + (x'' \mid a)\frac{\tau^2}{2!} + (x''' \mid a)\frac{\tau^3}{3!} + \cdots.$$

Hence, D is in general of the first order with respect to the arc PP'. It will be of at least the second order if and only if $(x' \mid a) = 0$; and at least of the third order if and only if $(x' \mid a) = 0$ and $(x'' \mid a) = 0$.

We proceed to interpret these results geometrically. The vanishing of $(x' \mid a)$ is the condition that the vector a normal to the given plane be perpendicular to the tangent line at P and hence that the plane contain this line. Consequently, the planes through the tangent line at P have contact of at least the first order with the curve and are the only planes with this property.

The equations $(x' \mid a) = 0$, $(x'' \mid a) = 0$ require that the vector a be perpendicular to the vector x'' as well as to the vector x'. In other words, they require that the vector a be parallel to the proper vector $\widehat{x' x''}$, and hence that the given plane be perpendicular to this vector. Thus, there is a unique plane through a regular point P of a curve, not a straight line, which has contact of at least the second order with the curve, namely, the plane through P which is perpendicular to the vector $\widehat{x'\,x''}$.

This plane is known as the *osculating plane* of the curve at P. Its equation is

$$(23)\qquad (X - x|\widehat{x'\,x''}) = 0 \quad \text{or} \quad (X - x\ x'\,x'') = 0,$$

where (X_1, X_2, X_3) are " running " coordinates.

The fact that the osculating plane at P is the plane through P which has the highest contact with the curve may be expressed roughly by saying that it is the plane through P to which the curve adheres most closely. The osculating plane of a plane curve is always the plane of the curve.

Example. In the case of the twisted cubic (4), the vectors x' and x'' have the components a, $2\,bt$, $3\,ct^2$ and 0, $2\,b$, $6\,ct$, and hence $\widehat{x'\,x''}$ has the components $6\,bct^2$, $-6\,act$, $2\,ab$. The equation of the osculating plane at the point with coordinate t is, then,

$$6\,bct^2(X_1 - at) - 6\,act(X_2 - bt^2) + 2\,ab(X_3 - ct^3) = 0,$$

or

$$3\,bct^2X_1 - 3\,actX_2 + abX_3 - abct^3 = 0.$$

EXERCISE

The osculating plane at a regular point P has contact of order n with the curve when and only when at P the vectors $x', x'', \cdots, x^{[n]}$ are coplanar, whereas $x^{[n+1]}$ does not lie in their plane.

15. Trihedral at a point. The plane through the point P which is perpendicular to the tangent line at P is called the *normal plane* of the curve at P, and the lines in it which go through P are known as the normals to the curve at P. Among the normals at a regular point P of a curve, not a straight line, there are two of fundamental importance: the *principal normal*, the normal lying in the osculating plane; and the *binormal*, the normal perpendicular to the osculating plane.

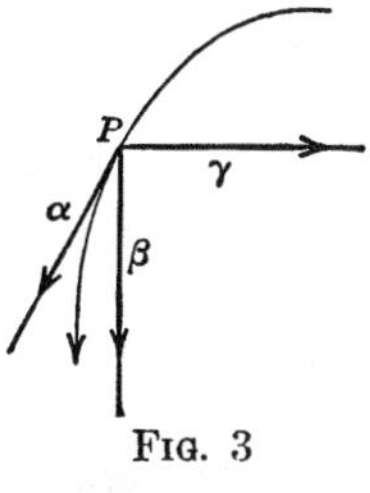

FIG. 3

The tangent, principal normal, and binormal are mutually perpendicular. They constitute what is called the trihedral at the point P.

Direction cosines of the directed tangent are

$$\alpha = \frac{x'}{\sqrt{x' \,|\, x'}}.$$

According to § 14, the binormal has the direction components $\widehat{x'\,x''}$. Hence

$$\gamma = \frac{\widehat{x'\,x''}}{\sqrt{x'\,x'' \,|\, x'\,x''}}$$

represents direction cosines, γ_1, γ_2, γ_3, of the binormal, directed in a definite sense.

We now choose the positive sense on the principal normal so that the directed tangent, the directed principal normal, and the directed binormal have, in the order given, the same disposition as the axes. If β_1, β_2, β_3 are the direction cosines of the principal normal, thus directed, it follows, by § 5, that $\beta = \widehat{\gamma\,\alpha}$, and hence

$$\beta = \frac{\widehat{\widehat{x'\,x''}\,x'}}{\sqrt{x' \,|\, x'}\,\sqrt{x'\,x'' \,|\, x'\,x''}}.$$

We may rewrite our results in the form of ω-identities:

$$(24) \qquad (\alpha \,|\, \omega) \equiv \frac{(x' \,|\, \omega)}{\sqrt{x' \,|\, x'}}, \quad (\beta \,|\, \omega) \equiv \frac{(x'\,x'' \,|\, x'\,\omega)}{\sqrt{x' \,|\, x'}\,\sqrt{x'\,x'' \,|\, x'\,x''}},$$

$$(\gamma \,|\, \omega) \equiv \frac{(x'\,x''\,\omega)}{\sqrt{x'\,x'' \,|\, x'\,x''}}.$$

These formulas give the direction cosines of the directed lines of our trihedral in terms of derivatives with respect to an arbitrary coordinate, or parameter, on the curve. The corresponding formulas in terms of the derivatives with respect to the arc s reduce, by means of the identities at the end of § 11, to the simple forms

$$(25) \qquad \alpha = x'_s, \qquad \beta = \frac{x''_s}{\sqrt{x''_s \,|\, x''_s}}, \qquad \gamma = \frac{\widehat{x'_s\,x''_s}}{\sqrt{x''_s \,|\, x''_s}}.$$

The vectors α, β, γ at P are the unit vectors in the positive directions of the directed edges of the trihedral. We have already called α the tangent vector at P. Similarly, we agree to call β the *principal normal vector* at P, and γ the *binormal vector*.

Planes of the trihedral. The equations of the planes determined by the edges of the trihedral, taken in pairs, are

$$(26)\qquad (X - x\,|\,\alpha) = 0, \quad (X - x\,|\,\beta) = 0, \quad (X - x\,|\,\gamma) = 0.$$

The first of these is the normal plane at P, and the third, the osculating plane. The second, the plane of the tangent and binormal, is known as the *rectifying plane*, for reasons which will be given later.

Example. For the circular helix

$$(27)\qquad x_1 = a\cos t, \quad x_2 = a\sin t, \quad x_3 = a\cot\theta\, t, \quad \cot\theta \neq 0,$$

the components of the vectors x', x'', x''' are, respectively,

$$-a\sin t,\ a\cos t,\ a\cot\theta; \quad -a\cos t,\ -a\sin t,\ 0; \quad a\sin t,\ -a\cos t,\ 0.$$

Those of $\widehat{x'\,x''}$ and $\widehat{\widehat{x'\,x''}\,x'}$ are then found to be

$$a^2\cot\theta\sin t,\ -a^2\cot\theta\cos t,\ a^2; \quad -a^3\csc^2\theta\cos t,\ -a^3\csc^2\theta\sin t,\ 0.$$

Moreover,

$$(x'\,|\,x') = a^2\csc^2\theta, \quad (x'\,x''\,|\,x'\,x'') = a^4\csc^2\theta, \quad (x'\,x''\,x''') = a^3\cot\theta,$$

where $(x'\,x''\,x''')$ has been computed for later purposes.

We may now find, by (24), the direction cosines of the vectors of the trihedral at an arbitrary point P. They are:

$$\begin{array}{llll}
\alpha: & -\sin\theta\sin t, & \sin\theta\cos t, & \cos\theta; \\
\beta: & -\cos t, & -\sin t, & 0; \\
\gamma: & \cos\theta\sin t, & -\cos\theta\cos t, & \sin\theta.
\end{array}$$

Since $\beta_3 = 0$, it follows that the principal normal to the circular helix at P is the line through P perpendicular to the axis of the cylinder (the x_3-axis).

16. Curvature. Osculating circle. The curvature of a curve at a point P, regular or singular, is the limit, when Δs approaches zero, of the numerical value of the ratio $A/\Delta s$, where A is the angle between the directed tangents at P and a neighboring point P', and Δs is the arc PP'. Thus, if the curvature is denoted by $1/R$,

$$(28)\qquad \frac{1}{R} = \lim_{\Delta s \to 0} \left| \frac{A}{\Delta s} \right|.$$

Since the tangent vectors at P and P' are α and $\alpha + \Delta\alpha$,

$$\cos A = (\alpha | \alpha + \Delta\alpha) = (\alpha | \alpha) + (\alpha | \Delta\alpha).$$

Now, $(\alpha|\alpha) = 1$; moreover, $(\alpha + \Delta\alpha | \alpha + \Delta\alpha) = 1$ and hence $2(\alpha|\Delta\alpha) + (\Delta\alpha|\Delta\alpha) = 0$ or $(\alpha|\Delta\alpha) = -(1/2)(\Delta\alpha|\Delta\alpha)$. Thus,

$$\cos A = 1 - \tfrac{1}{2}(\Delta\alpha|\Delta\alpha) \quad \text{or} \quad 2(1 - \cos A) = (\Delta\alpha|\Delta\alpha),$$

and

$$\frac{2(1 - \cos A)}{A^2}\,\frac{A^2}{\Delta s^2} = \left(\frac{\Delta\alpha}{\Delta s}\middle|\frac{\Delta\alpha}{\Delta s}\right).$$

But

$$\lim_{\Delta s \to 0} \frac{2(1 - \cos A)}{A^2} = 1, \quad \lim_{\Delta s \to 0} \frac{A^2}{\Delta s^2} = \frac{1}{R^2}, \quad \lim_{\Delta s \to 0} \frac{\Delta\alpha}{\Delta s} = \frac{d\alpha}{ds}.$$

Hence

$$\frac{1}{R^2} = \left(\frac{d\alpha}{ds}\middle|\frac{d\alpha}{ds}\right),$$

or, since $\alpha = x'_s$,

$$\text{(29)} \qquad \frac{1}{R} = \sqrt{x''_s | x''_s}.$$

THEOREM 1. *A necessary and sufficient condition that the curvature of a curve be identically zero is that the curve be a straight line.*

According to § 13, Theorem 1, a curve which is represented parametrically in terms of its arc s is a straight line if and only if $(x'_s x''_s | x'_s x''_s) \equiv 0$. But, by (18), $(x'_s x''_s | x'_s x''_s) \equiv (x''_s | x''_s)$, and the theorem follows.

A similar argument tells us that *the curvature of a curve, not a straight line, is zero at a point P when and only when P is a singular point.*

At a regular point P of a curve, not a straight line, $\beta = x''_s / \sqrt{x''_s | x''_s}$. But $x''_s = d\alpha/ds$ and $\sqrt{x''_s | x''_s} = 1/R$. Hence

$$\text{(30)} \qquad \frac{d\alpha}{ds} = \frac{\beta}{R}.$$

Osculating circle. We proceed to find, if it exists, the circle tangent to the curve at P which has contact of the highest order with the curve. Let C be an arbitrarily chosen, but fixed, circle tangent to the curve at P, and let M be its center. Mark the foot, F, of the perpendicular dropped from P' on the plane of C, and let Q be that point of intersection of the line FM with C which is nearer P'. Then $P'Q$ is the perpendicular distance from P' to C. Evidently (Fig. 4),

$$\overline{P'Q}^2 = \overline{P'F}^2 + \overline{QF}^2. \tag{31}$$

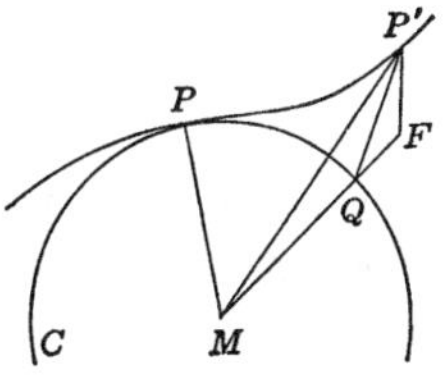

FIG. 4

The infinitesimal $P'F$, since it is the distance from the point P' to a plane through the tangent line at P, is of the second order unless the plane is the osculating plane, when it is of at least the third order; see § 14. Consequently, the infinitesimal $P'Q$ in which we are interested cannot be of order higher than the second unless the circle C lies in the osculating plane.

In case C does lie in the osculating plane, its center M is on the principal normal to the curve at P and hence has the coordinates

$$z = x + r\beta,$$

where r is the directed distance from P to M (see § 7) and $|r|$ is the radius of C.

Since we know that $P'F$ is now of at least the third order, it remains merely to try to choose r, and hence C, so that QF is also of at least the third order. If this is possible and the choice is unique, there will be a single circle for which the infinitesimal $P'Q$ is at least of the third order.

From the right triangle MFP',

$$(QF + |r|)^2 = \overline{MP'}^2 - \overline{P'F}^2.$$

Hence

$$(2|r| + QF)QF = \overline{MP'}^2 - r^2 - \overline{P'F}^2. \tag{32}$$

Since M has the coordinates $z = x + r\beta$ and P' has the coordinates

$$y = x + x_s'\sigma + x_s''\frac{\sigma^2}{2!} + x_s'''\frac{\sigma^3}{3!} + \cdots,$$

we have

$$y - z = -r\beta + x_s'\sigma + x_s''\frac{\sigma^2}{2!} + x_s'''\frac{\sigma^3}{3!} + \cdots,$$

whence

$$\overline{MP'}^2 = r^2 + (1 - r(\beta | x_s''))\sigma^2 + \cdots .$$

It follows then, from (32), since $\overline{P'F}^2$ is of at least the sixth order, that

$$(2|r| + QF)\, QF = (1 - r(\beta | x_s''))\sigma^2 + \cdots .$$

Hence there is a unique value of r for which QF is of at least the third order, namely that defined by the equation $1 - r(\beta | x_s'') = 0$. Moreover, since $x_s'' = \beta/R$, this value of r is precisely R.

THEOREM 2. *There is a unique circle which has contact of at least the second order with a given curve, not a straight line, at a regular point. It is the circle in the osculating plane whose radius is R and whose center lies on the positive half of the principal normal, in the point $x + R\beta$.*

This circle is known as *the osculating circle*, or *the circle of curvature*, of the curve at the point. Its center is called the *center of curvature*, and its radius, the *radius of curvature*, for the point.

17. Torsion. The curvature of a curve at a point P is a measure of the rate at which the curve is turning away from the tangent line at P. The torsion of the curve at P is a similar measure of the rate at which the curve is twisting out of the osculating plane at P. At a regular point P of a curve, not a straight line, it is, to within sign, the limit, when Δs approaches zero, of the ratio $C/\Delta s$, where C is the angle between the osculating planes at P and a neighboring point P', and Δs is the arc PP'.

The angle C may be better described as the angle between the directed binormals at P and P', that is, as the angle between the unit vectors γ and $\gamma + \Delta\gamma$. Hence, proceeding as in the case of the curvature, we find

$$\frac{1}{T^2} = \left(\frac{d\gamma}{ds}\middle|\frac{d\gamma}{ds}\right). \tag{33}$$

Since

$$\gamma = \frac{\widehat{x_s' x_s''}}{\sqrt{x_s'' | x_s''}},$$

$$\frac{d\gamma}{ds} = \frac{(x_s'' | x_s'')\widehat{x_s' x_s'''} - (x_s'' | x_s''')\widehat{x_s' x_s''}}{(x_s'' | x_s'')^{3/2}}. \tag{34}$$

If in the identity (24) of Chapter I, namely,

$$(a\,b\,c)(d\,|\,e) = (b\,c\,d)(a\,|\,e) + (c\,a\,d)(b\,|\,e) + (a\,b\,d)(c\,|\,e),$$

we set $a = x_s'$, $b = x_s''$, $c = x_s'''$, $d = \omega$, and $e = x_s''$, we get, since $(x_s'\,|\,x_s'') = 0$,

$$(x_s'\,x_s''\,x_s''')(x_s''\,|\,\omega) = -(x_s''\,|\,x_s'')(x_s'\,x_s'''\,\omega) + (x_s''\,|\,x_s''')(x_s'\,x_s''\,\omega).$$

From this identity in the triple ω, it is evident that the numerator on the right-hand side of (34) has the value $-(x_s'\,x_s''\,x_s''')x_s''$. Thus

$$\frac{d\gamma}{ds} = -\frac{(x_s'\,x_s''\,x_s''')}{(x_s''\,|\,x_s'')^{3/2}}\,x_s''. \tag{35}$$

Hence

$$\frac{1}{T^2} = \frac{(x_s'\,x_s''\,x_s''')^2}{(x_s''\,|\,x_s'')^2} \quad \text{and} \quad \frac{1}{T} = \pm\frac{(x_s'\,x_s''\,x_s''')}{(x_s''\,|\,x_s'')}\,.$$

We choose the minus sign in the later formula, thus completing our definition of the torsion. We have, then,

$$\frac{1}{T} = -\frac{(x_s'\,x_s''\,x_s''')}{(x_s''\,|\,x_s'')}\,. \tag{36}$$

We may now rewrite (35), since $x_s''/\sqrt{x_s''\,|\,x_s''} = \beta$, in the form

$$\frac{d\gamma}{ds} = \frac{\beta}{T}\,. \tag{37}$$

We give a second deduction of formulas (36) and (37), based on the fact that α, β, γ are mutually perpendicular unit vectors. From $(\gamma\,|\,\alpha) = 0$, we obtain, by differentiation, the equation $(d\gamma/ds\,|\,\alpha) + (\gamma\,|\,d\alpha/ds) = 0$, which, since $d\alpha/ds = \beta/R$, reduces to

$$\left(\frac{d\gamma}{ds}\,\middle|\,\alpha\right) = 0.$$

Since $(\gamma\,|\,\gamma) = 1$, we have also

$$\left(\frac{d\gamma}{ds}\,\middle|\,\gamma\right) = 0.$$

These two relations tell us that $d\gamma/ds = c\beta$. Hence $(d\gamma/ds\,|\,d\gamma/ds) = c^2$, so that, by (33), $c^2 = (1/T)^2$. Taking $c = 1/T$ and thus fixing the sign of the torsion, we obtain formula (37): $d\gamma/ds = \beta/T$. Forming the inner product of both sides of (37) with β, we obtain for $1/T$ the expression $1/T = (d\gamma/ds\,|\,\beta)$, which reduces to (36) when we set $\beta = x_s''/\sqrt{x_s''\,|\,x_s''}$ and use for $d\gamma/ds$ the value given in (34).

The formulas

$$\frac{d\alpha}{ds} = \frac{\beta}{R}, \qquad \frac{d\gamma}{ds} = \frac{\beta}{T} \tag{38}$$

associate the curvature and torsion of a curve with the rates of change, with respect to the arc, of the tangent and binormal vectors, respectively. May there, then, not be a third "curvature," associated with the rate of change of the principal normal vector β?

This question is evidently to be answered in the negative, inasmuch as β is expressible in terms of γ and α:

$$\beta = \widehat{\gamma\,\alpha}.$$

Hence

$$\frac{d\beta}{ds} = \widehat{\gamma\frac{d\alpha}{ds}} + \widehat{\frac{d\gamma}{ds}\alpha} = \frac{1}{R}\widehat{\gamma\,\beta} + \frac{1}{T}\widehat{\beta\,\alpha},$$

or, since $\widehat{\gamma\,\beta} = -\,\alpha$ and $\widehat{\beta\,\alpha} = -\,\gamma$ (§ 5),

$$\frac{d\beta}{ds} = -\,\frac{\alpha}{R} - \frac{\gamma}{T}. \tag{39}$$

The curvature and torsion have thus far been expressed only in terms of the derivatives with respect to the arc s:

$$\frac{1}{R} = \sqrt{x_s''\,|\,x_s''}, \qquad \frac{1}{T} = -\,\frac{(x_s'\,x_s''\,x_s''')}{(x_s''\,|\,x_s'')}. \tag{40}$$

The corresponding expressions in terms of derivatives with respect to the arbitrary parameter t are

$$\frac{1}{R} = \frac{\sqrt{\widehat{x'\,x''}\,|\,\widehat{x'\,x''}}}{(x'\,|\,x')^{3/2}}, \qquad \frac{1}{T} = -\,\frac{(x'\,x''\,x''')}{(\widehat{x'\,x''}\,|\,\widehat{x'\,x''})}. \tag{41}$$

These expressions are obtained from (40) by substituting for the derivatives with respect to s their values in terms of the derivatives with respect to t: $x_s' = x'/\sqrt{x'\,|\,x'}$, etc. The computation, though somewhat laborious, is perfectly straightforward.

Example. For the circular helix, we readily find, by use of (41) and the data in § 15, that

$$\frac{1}{R} = \frac{\sin^2 \theta}{a}, \qquad \frac{1}{T} = -\frac{\sin \theta \cos \theta}{a}. \tag{42}$$

Thus, the curvature and torsion of a circular helix are both constant.

18. Plane curves. If the curve $x = x(t)$ is a plane curve, not a straight line, the plane of the curve is the osculating plane at every point. The binormal vector γ, considered for the moment as a free vector, is, then, always the same, that is, $d\gamma/ds \equiv 0$. Hence, by (33), $1/T \equiv 0$.

Suppose, conversely, that $1/T \equiv 0$. Then $d\gamma/ds \equiv 0$. Moreover, $d(\gamma | x)/ds \equiv (\gamma | \alpha) + (d\gamma/ds | x) \equiv 0$, and hence $(\gamma | x) \equiv c$. Since γ_1, γ_2, γ_3 and c are constants, this identity says that the curve $x = x(t)$ lies in the plane $(\gamma | X) = c$.

THEOREM 1. *A necessary and sufficient condition that a curve, not a straight line, be a plane curve is that its torsion be identically zero.*

The proof of the theorem is open to the objection that it does not cover the case of singular points; for, at these points the torsion and the osculating plane have not been defined. This objection will be met later.

THEOREM 2. *A necessary and sufficient condition that the curve* $x = x(t)$ *be a plane curve is that* $(x' \, x'' \, x''') \equiv 0$.

The theorem follows from (41) and Theorem 1 in case the curve is not a straight line and is guaranteed by Theorem 1 of § 13 when the curve is a straight line.

Closely related to Theorem 2 is the following theorem.

THEOREM 3. *The vectors of the one-parameter family of proper vectors* $a = a(t)$ *are all parallel to a plane if and only if* $(a \, a' \, a'') \equiv 0$.

If the vectors a are all parallel to a plane, they are all perpendicular to a constant proper vector c. Hence

$$(a | c) \equiv 0, \qquad (a' | c) \equiv 0, \qquad (a'' | c) \equiv 0,$$

the last two equations being obtained from the first by successive differentiations. By hypothesis, the three equations have a solution c_1, c_2, c_3, other than 0, 0, 0, for every value of t under consideration. Hence, the determinant of the coefficients of c_1, c_2, c_3 vanishes identically, that is, $(a\,a'\,a'') \equiv 0$.

Suppose, conversely, that $(a\,a'\,a'') \equiv 0$ and consider the curve $x = \int a(t)dt$. Since $x' \equiv a$, the vectors $a(t)$ may be thought of as vectors tangential to this curve. It will follow that they are parallel to a plane if we can show that the curve is a plane curve. But, inasmuch as $x' \equiv a$, we have $x'' \equiv a'$, $x''' \equiv a''$ and hence $(x'\,x''\,x''') \equiv (a\,a'\,a'') \equiv 0$. Thus, the curve is a plane curve and the proof is complete.

EXERCISES

Find α, β, γ, $1/R$, $1/T$ for each of the following curves.

1. The twisted cubic (4).

2. The curve of § 9, Ex. 2. Show that the curve cuts itself at right angles. Determine the points at which the torsion is zero.

3. The curve of § 9, Ex. 3. Find the points at which the torsion is zero and identify them geometrically.

4. Show that the locus of the center of curvature of a circular helix, C, is a circular helix, C'. Describe the position of C' with respect to C. Show that at corresponding points of C and C' the tangents are mutually perpendicular and the curvatures equal. Determine the locus of the center of curvature of C'. When do C and C' lie on the same circular cylinder?

5. The plane curve $y = f(x)$ has the parametric representation $x_1 = x$, $x_2 = y = f(x)$, $x_3 = 0$, in terms of x as parameter. Find α, β, γ and deduce the formulas

$$\frac{1}{R} = \frac{|y''|}{\sqrt{(1 + y'^2)^3}}, \qquad s = \int_{x_0}^{x} \sqrt{1 + y'^2}\,dx.$$

Verify the fact that $1/T \equiv 0$.

6. Prove that, if the equations $x_1 = at$, $x_2 = bt^2$, $x_3 = f(t)$, $ab \neq 0$, represent a plane curve, the curve is a parabola.

7. Deduce formulas (41) from (40).

19. The Frenet-Serret formulas. Formulas (38) and (39) of § 17, namely,

$$\frac{d\alpha}{ds} = \frac{\beta}{R}, \qquad \frac{d\beta}{ds} = -\frac{\alpha}{R} - \frac{\gamma}{T}, \qquad \frac{d\gamma}{ds} = \frac{\beta}{T}, \tag{43}$$

are the central formulas in the theory of a space curve. They were discovered independently, about 1850, by two French mathematicians, Frenet and Serret.

By means of them, it is possible to show that the vectors x_s', x_s'', x_s''', $\cdots$ may all be expressed as linear combinations of α, β, γ in which the scalars of combination are rational functions of $1/R$, $1/T$, and the derivatives of $1/R$ and $1/T$ with respect to s. For example,

$$x_s' = \alpha, \qquad x_s'' = \frac{\beta}{R},$$
$$x_s''' = \frac{1}{R}\left(-\frac{\alpha}{R} - \frac{\gamma}{T}\right) + \frac{d}{ds}\left(\frac{1}{R}\right)\beta. \tag{44}$$

Let us now fix our attention on a regular point P_0 of the curve, and measure the arc s from this point. Then the neighboring point P for which the arc is s has, by virtue of relations (44), the coordinates

$$y = x + \alpha s + \frac{\beta}{R_0}\frac{s^2}{2!} + \left(-\frac{1}{R_0^2}\alpha + \frac{d}{ds}\left(\frac{1}{R}\right)_0\beta - \frac{1}{R_0T_0}\gamma\right)\frac{s^3}{3!} + \cdots, \tag{45}$$

where (x_1, x_2, x_3) are the coordinates of P_0, and α, β, γ are the vectors of the trihedral at P_0.

Suppose we assume, further, that the original coordinate system in space was chosen so that P_0 is the origin and the directed edges of the trihedral at P_0, in the usual order, are respectively the positive axes of x_1, x_2, x_3. Then P_0 is the point (0, 0, 0), and α, β, γ have respectively the components 1, 0, 0, 0, 1, 0, 0, 0, 1. Hence the equations for which (45) stands become

$$\begin{aligned} y_1 &= s - \frac{1}{6}\frac{1}{R_0^2}s^3 + \cdots, \\ y_2 &= \frac{1}{2R_0}s^2 + \frac{1}{6}\frac{d}{ds}\left(\frac{1}{R}\right)_0 s^3 + \cdots, \\ y_3 &= -\frac{1}{6}\frac{1}{R_0T_0}s^3 + \cdots. \end{aligned} \tag{46}$$

These equations constitute what is called the *canonical representation* of the curve in *the* neighborhood of the point P_0 (see § 8). It is to be noted that all the coefficients in the three power series depend simply on the values, at P_0, of $1/R$, $1/T$, and the derivatives of $1/R$, $1/T$ with respect to s.

From equations (46) we derive various properties of the curve in the neighborhood of the point P_0. For example, since y_3 is the (directed) distance from the osculating plane at P_0 to the point P, we deduce the following fact.

THEOREM 1. *If the torsion at a regular point P_0 is not zero, the osculating plane at P_0 has precisely contact of the second order with the curve and crosses the curve at P_0.*

Since $1/R_0 \neq 0$, the sign of the coefficient of s^3 in the expression for y_3 depends simply on the sign of $1/T_0$. Thus we obtain the following interpretation of the sign of the torsion.

COROLLARY. *If the curve in the neighborhood of P_0 is traversed in the positive direction, it pierces the osculating plane at P_0 from the side of the positive or negative binormal according as the torsion at P_0 is positive or negative.*

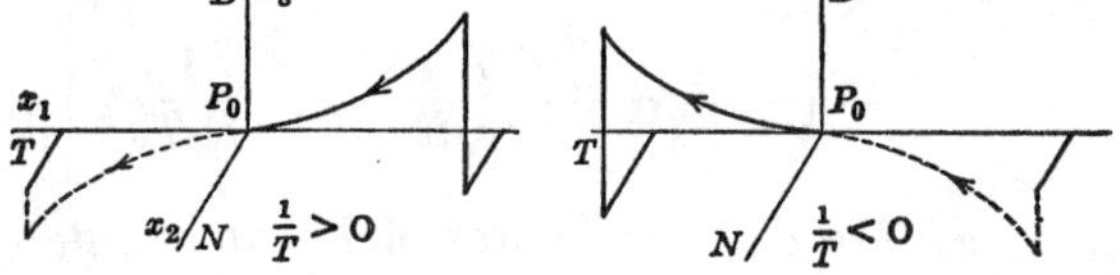

FIG. 5

When s is sufficiently small numerically, it is evident that $y_2 > 0$ except when $s = 0$. Hence, in a sufficiently restricted neighborhood of P_0, the curve lies on one side of the rectifying plane, namely, on the side toward the positive principal normal. This is in agreement with the fact that the center of curvature for P_0 lies on the positive half of the principal normal.

We consider finally the projections, on the planes of the trihedral at P_0, of the curve in the neighborhood of P_0. These projections are represented parametrically, in their several planes, by the three pairs of equations obtained by suppressing each of the equations (46) in turn.

In these pairs of equations we proceed to discard all the terms in each series except the leading term. Thereby, we obtain parametric representations, not of the projections, but of curves which approximate to the projections. These curves are:

In the osculating plane: $x_1 = s, \qquad x_2 = \frac{1}{2R_0} s_2.$

In the normal plane: $x_2 = \dfrac{1}{2R_0}s^2, \qquad x_3 = -\dfrac{1}{6R_0T_0}s^3.$

In the rectifying plane: $x_3 = -\dfrac{1}{6R_0T_0}s^3, \qquad x_1 = s.$

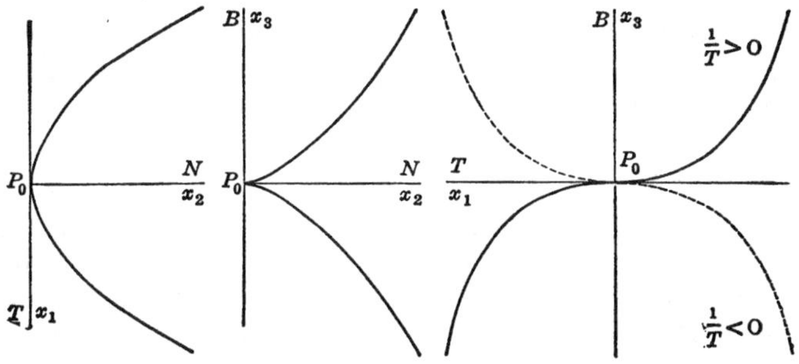

FIG. 6

Consequently, to within the degree of approximation described, the three projections are respectively a parabola ($x_2 = ax_1^2$), a semi-cubical parabola ($x_3^2 = bx_2^3$), and a cubic curve ($x_3 = cx_1^3$). See Fig. 6.

EXERCISE

Show that the curvature of a twisted curve at a point P is equal to the curvature at P of the plane curve which is the projection of the twisted curve on the osculating plane at P.

20. Singular points. At a singular point P_0 of the curve $x = x(t)$, that is, a point for which $\widehat{x'x''} = 0$ or $(x'\,x''|x'\,x'') = 0$, the curvature is zero. The torsion, osculating plane, principal normal, and binormal are still to be defined.

Since an analytic function of a single variable, such as $(x'\,x''|x'\,x'')$, can vanish only for isolated values of the variable, there exists a neighborhood of the curve about P_0 in which P_0 is the only singular point. For every point P, except P_0, in this neighborhood, the torsion $1/T$ is defined and analytic. It is natural, then, to define the torsion at P_0 as the limit of $1/T$ when P approaches P_0. It can be shown that this limit actually exists and that, when we take it as the value of $1/T$ at P_0, $1/T$ is not only continuous at P_0, but analytic there.

Similar remarks apply to the osculating plane and to the principal normal and binormal. At every point P, except P_0, in the neighborhood of P_0 in question, the osculating plane is defined and changes according to a law which is analytic in the sense in which we have been using the term. As the osculating plane at P_0, we take the limit of the osculating plane at P when P approaches P_0. Thereby, the principal normal and binormal vectors at P_0, and their components, are defined as the limits, when P approaches P_0, of the principal normal and binormal vectors at P, and their components. Then β_1, β_2, β_3 and γ_1, γ_2, γ_3 are not only continuous at P_0, but also analytic.

When β_i, γ_i $(i = 1, 2, 3)$ and $1/T$ have been defined at each singular point in the manner described, the functions α_i, β_i, γ_i $(i = 1, 2, 3)$ and $1/R$, $1/T$ are analytic at every point of the curve. Moreover, since the definitions in question are the result of taking limits, and the limits of equals are equal, relations between these functions which have been established for regular points are now valid also at singular points. Hence, the Frenet-Serret formulas hold for all points of the curve. Incidentally, the proof of Theorem 1 of § 18 is thus raised above criticism.

The result of neglecting terms of the third and higher orders in equations (46) is $y_1 = s$, $y_2 = s^2/(2R_0)$, $y_3 = 0$. Thus, to within infinitesimals of at least the third order the curve is, in a restricted neighborhood of a regular point, a parabola, but in a restricted neighborhood of a singular point, a straight line. Hence, the curve behaves to a certain degree like a straight line at a singular point, and, what is equally important, is perfectly smooth there.

At a point P_0 of a twisted curve, the vectors x', x'', x''', x^{iv}, $\cdots$ can be arranged in three groups: x', x'', $\cdots$, $x^{[m]}$; $x^{[m+1]}$, $\cdots$, $x^{[m+n]}$; $x^{[m+n+1]}$, $\cdots$, so that (a) the vectors x', x'', $\cdots$, $x^{[m]}$ are collinear, but $x^{[m+1]}$ does not lie on their line, and (b) the vectors x', x'', $\cdots$, $x^{[m+n]}$ are coplanar, whereas $x^{[m+n+1]}$ does not lie in their plane. It follows, then, that the tangent line at P_0 has contact of order m with the curve, and the osculating plane, contact of order $m + n$. Moreover, it can be shown, though not without some computation, that the first terms in the power series developments of $1/R$ and $1/T$ at P_0, expressed in terms of the arc s measured from P_0, are respectively as^{m-1} $(a \neq 0)$ and cs^{n-1} $(c \neq 0)$.

When $m = 1$ and $n = 1$, P_0 is a regular point at which $1/T \neq 0$. When $m = 1$ and $n > 1$, P_0 is a regular point at which the curve behaves like a plane curve $(1/T = 0)$. When $m > 1$ and $n = 1$,

P_0 is a singular point at which the curve behaves like a straight line ($1/R = 0$), but not like a plane curve! Finally, when $m > 1$ and $n > 1$, P_0 is a singular point at which the curve behaves both like a straight line and a plane curve.

That a curve can behave at a point like a straight line but not like a plane curve, that is, that the curvature may vanish without the torsion vanishing also, simply reflects the fact that the curvature and torsion of a curve are independent.

The singular points of a regular analytic curve are, as we have seen, so innocuous as hardly to deserve the name. On the other hand, at points of an analytic curve at which the curve fails to be regular in that $x' = 0$, the curve may turn sharply on itself in various sorts of cusps and beaks.

21. Fundamental theorem. Inasmuch as we are interested only in properties of a curve which are preserved by every rigid motion, it is of first importance to have a criterion for the congruence of two curves in terms of properties of the curves which are independent of their positions in space. A criterion of this sort is given by the following theorem.

THEOREM 1. *A necessary and sufficient condition that two curves, C and $\overline{C}$, be congruent is that the points of one may be put into a one-to-one correspondence with the points of the other so that corresponding arc lengths, and the curvatures and torsions at corresponding points, are equal.*

Since length of arc, curvature, and torsion of a curve are preserved by a rigid motion, the condition is necessary.

In the proof of the sufficiency, we shall need the following lemma.

LEMMA. *If the number triples α, β, γ satisfy the equations*

$$(\alpha|\alpha) = (\beta|\beta) = (\gamma|\gamma) = 1, \quad (\beta|\gamma) = (\gamma|\alpha) = (\alpha|\beta) = 0, \tag{47}$$

then they also satisfy the equations

$$\begin{aligned} \alpha_i^2 + \beta_i^2 + \gamma_i^2 &= 1, & i &= 1, 2, 3, \\ \alpha_i\alpha_j + \beta_i\beta_j + \gamma_i\gamma_j &= 0, & i \neq j;\ i, j &= 1, 2, 3. \end{aligned} \tag{48}$$

Conversely, if the relations (48) *hold, relations* (47) *follow.*

Relations (47) say that $\alpha_1, \alpha_2, \alpha_3$, $\beta_1, \beta_2, \beta_3$, $\gamma_1, \gamma_2, \gamma_3$ are respectively the direction cosines, with respect to the axes of x_1, x_2,

x_3, of three mutually perpendicular directed lines A, B, C, through a point P. But, then, $\alpha_1, \beta_1, \gamma_1, \alpha_2, \beta_2, \gamma_2, \alpha_3, \beta_3, \gamma_3$ are respectively the direction cosines, with respect to the directed lines A, B, C, of the axes of x_1, x_2, x_3. Consequently, relations (48) are valid. The converse is proved in the same way.

To prove the sufficiency of Theorem 1, we have to show that, if there exists between the points of C and $\bar{C}$ a one-to-one correspondence of the type described, then C and $\bar{C}$ are congruent. Let P_0 be a fixed point of C, $\bar{P}_0$ the corresponding point of $\bar{C}$, and visualize the trihedral of C at P_0 and the trihedral of $\bar{C}$ at $\bar{P}_0$. If there exists a rigid motion which carries the points of $\bar{C}$ into the corresponding points of C, it must be the rigid motion which carries $\bar{P}_0$ into P_0 and the directed edges of the trihedral at $\bar{P}_0$ into the corresponding directed edges of the trihedral at P_0. We assume that this rigid motion has been carried out and proceed to prove that $\bar{C}$, in its new position, is identical with C.

Think of the arcs of C and $\bar{C}$ both measured from the point P_0 in corresponding directions. Then, if P and $\bar{P}$ are an arbitrary pair of corresponding points, the arcs P_0P and $P_0\bar{P}$ have the same value, s. Moreover, since the curvatures and torsions of C and $\bar{C}$ at P and $\bar{P}$ are equal, they are the same functions of s for both curves. Hence, if α, β, γ and $\bar{\alpha}, \bar{\beta}, \bar{\gamma}$ are the vectors of the trihedrals of C and $\bar{C}$ at P and $\bar{P}$, the Frenet-Serret formulas of the two curves read

$$\begin{aligned} \frac{d\alpha_i}{ds} &= \frac{\beta_i}{R}, \quad \frac{d\beta_i}{ds} = -\frac{\alpha_i}{R} - \frac{\gamma_i}{T}, \quad \frac{d\gamma_i}{ds} = \frac{\beta_i}{T}, \\ \frac{d\bar{\alpha}_i}{ds} &= \frac{\bar{\beta}_i}{R}, \quad \frac{d\bar{\beta}_i}{ds} = -\frac{\bar{\alpha}_i}{R} - \frac{\bar{\gamma}_i}{T}, \quad \frac{d\bar{\gamma}_i}{ds} = \frac{\bar{\beta}_i}{T}, \end{aligned} \qquad i = 1, 2, 3.$$

By means of these formulas, it is readily shown that

$$\alpha_i \frac{d\bar{\alpha}_i}{ds} + \bar{\alpha}_i \frac{d\alpha_i}{ds} + \beta_i \frac{d\bar{\beta}_i}{ds} + \bar{\beta}_i \frac{d\beta_i}{ds} + \gamma_i \frac{d\bar{\gamma}_i}{ds} + \bar{\gamma}_i \frac{d\gamma_i}{ds} = 0, i = 1, 2, 3.$$

Consequently,

$$\text{(49)} \qquad \alpha_i\bar{\alpha}_i + \beta_i\bar{\beta}_i + \gamma_i\bar{\gamma}_i = c_i, \qquad i = 1, 2, 3,$$

where c_1, c_2, c_3 are constants.

These constants are all equal to unity. For, when $s = 0$, then $\bar{\alpha} = \alpha$, $\bar{\beta} = \beta$, $\bar{\gamma} = \gamma$ and (49) reduces to $\alpha_i^2 + \beta_i^2 + \gamma_i^2 = c_i$. But, by the Lemma, $\alpha_i^2 + \beta_i^2 + \gamma_i^2 = 1$, and hence $c_i = 1$.

We now have the identities

$$\alpha_i\bar{\alpha}_i + \beta_i\bar{\beta}_i + \gamma_i\bar{\gamma}_i \equiv 1, \qquad i = 1, 2, 3.$$

But, by the Lemma,

$$\alpha_i^2 + \beta_i^2 + \gamma_i^2 \equiv 1, \quad \bar{\alpha}_i^2 + \bar{\beta}_i^2 + \bar{\gamma}_i^2 \equiv 1, \quad i = 1, 2, 3.$$

These identities, for an arbitrary but fixed value of i, say that the two vectors with the components α_i, β_i, γ_i and $\bar{\alpha}_i$, $\bar{\beta}_i$, $\bar{\gamma}_i$ are unit vectors which make with one another the angle zero. Hence they are the same vector:

$$\bar{\alpha}_i \equiv \alpha_i, \qquad \bar{\beta}_i \equiv \beta_i, \qquad \bar{\gamma}_i \equiv \gamma_i, \qquad i = 1, 2, 3.$$

Since $\bar{\alpha}_i \equiv \alpha_i$, we have

$$\frac{d}{ds}(\bar{x}_i - x_i) \equiv 0, \qquad i = 1, 2, 3,$$

whence

$$\bar{x}_i - x_i \equiv k_i, \qquad i = 1, 2, 3.$$

But, for $s = 0$, $\bar{x}_i = x_i$ and so k_1, k_2, k_3 are all zero. Consequently, $\bar{x}_i \equiv x_i$, $i = 1, 2, 3$, and $\bar{C}$ coincides with C.

Our theorem says that a given curve is uniquely determined, except for its position in space, by its curvature and torsion expressed as functions of the arc. Hence, if these functions are respectively $f(s)$ and $\phi(s)$, the equations

$$\frac{1}{R} = f(s), \qquad \frac{1}{T} = \phi(s) \tag{50}$$

define the curve completely, except for its position.

Since the curvature, torsion, and arc are natural or inherent properties of the curve, equations (50) are called *intrinsic* or *natural equations* of the curve.

The natural equations of a circular helix are, according to (42), of the form $1/R = c$, $1/T = k$, where c is a positive constant and k is a constant, not zero. Conversely, every pair of equations of this form represents a circular helix; for, the constants a and θ in (42) may be so chosen that $1/R$ and $1/T$ have the prescribed values c and k.

A plane curve has the single intrinsic equation $1/R = f(s)$. If the curve is a circle, this equation is $1/R = c$ $(c > 0)$, and conversely.

The natural equation of a plane curve may be found by eliminating the parameter employed to represent the curve from the equations expressing the curvature and arc (measured from a specific point) in terms of this parameter. The same method may be employed in the case of a space curve. Of course, the elimination, in either case, may present difficulties.

Example 1. For the catenary, $y = (a/2)(e^{x/a} + e^{-x/a})$,

$$4\,R = a(e^{x/a} + e^{-x/a})^2, \qquad 2\,s = a(e^{x/a} - e^{-x/a}),$$

where s is measured from the vertex. Hence, the natural equation is found to be $aR = s^2 + a^2$.

Example 2. As an example of a curve whose equation is most simply expressed in terms of polar coordinates, we take the logarithmic spiral: $r = ce^{\cot\alpha\,\theta}$, where c is a positive constant and α is a fixed acute angle. If s is measured from the point $(c, 0)$, we find that $R = \csc\alpha\, r$, $s = \sec\alpha\,(r - c)$, and hence $R = \cot\alpha\, s + c\csc\alpha$. It follows, then, that every equation of the form $R = as + b$, where $b > 0$ and $a \neq 0$, represents a logarithmic spiral.

Not only does every curve have intrinsic equations of the form (47), but every pair of equations of this form represents a curve, provided merely that the function $f(s)$ chosen is positive. In other words:

THEOREM 2. *If $f(s)$ and $\phi(s)$ are real, single valued, analytic functions of the real variable s in an interval including $s = 0$, and $f(s)$ is positive throughout this interval, there exists a curve whose curvature and torsion, expressed as functions of the arc measured from a suitably chosen point, are precisely $f(s)$ and $\phi(s)$.*

The theorem says, in brief, that the curvature and torsion may be prescribed at pleasure as functions of the arc.

The proof depends on the properties of the system of three ordinary differential equations,

$$(51) \qquad \frac{da}{ds} = f(s)b, \quad \frac{db}{ds} = -f(s)a - \phi(s)c, \quad \frac{dc}{ds} = \phi(s)b,$$

in the three unknown scalar functions $a(s)$, $b(s)$, $c(s)$.

It is known that, if a_0, b_0, c_0 are given constants, equations (51) have a unique solution $a(s)$, $b(s)$, $c(s)$, for which $a(0) = a_0$, $b(0) = b_0$, $c(0) = c_0$. Hence there exist unique solutions,

$$\alpha_1, \beta_1, \gamma_1; \qquad \alpha_2, \beta_2, \gamma_2; \qquad \alpha_3, \beta_3, \gamma_3, \tag{52}$$

which for $s = 0$ take on the prescribed values

$$1, 0, 0; \qquad 0, 1, 0; \qquad 0, 0, 1. \tag{53}$$

We proceed to show that these three solutions satisfy relations (48) identically. We know that

$$\begin{aligned} \frac{d\alpha_i}{ds} &= f(s)\beta_i, & \frac{d\beta_i}{ds} &= -f(s)\alpha_i - \phi(s)\gamma_i, & \frac{d\gamma_i}{ds} &= \phi(s)\beta_i, \\ \frac{d\alpha_j}{ds} &= f(s)\beta_j, & \frac{d\beta_j}{ds} &= -f(s)\alpha_j - \phi(s)\gamma_j, & \frac{d\gamma_j}{ds} &= \phi(s)\beta_j. \end{aligned} \tag{54}$$

Hence, by the method used in the proof of Theorem 1, we get

$$\alpha_i\alpha_j + \beta_i\beta_j + \gamma_i\gamma_j = c_{ij}.$$

Introducing the initial conditions, for $s = 0$, we find that $c_{ii} = 1$ and $c_{ij} = 0$, where $i \neq j$, and $i, j = 1, 2, 3$. Hence our statement is established.

Since the functions (52) satisfy relations (48) identically, they satisfy relations (47) identically. Consequently, the vectors α, β, γ with the components α_1, α_2, α_3, β_1, β_2, β_3, γ_1, γ_2, γ_3 are mutually perpendicular unit vectors which have, in the order given, the same disposition as the axes.

Consider, now, the curve

$$x = \int_0^s \alpha\, ds. \tag{55}$$

Its arc and tangent vector are evidently s and α. Hence it is readily shown, by comparing the Frenet-Serret formulas for it with equations (54), that its curvature and torsion are precisely $f(s)$ and $\phi(s)$. Thus, the theorem is proved.

Example 3. We shall now integrate equations (51) in the case of a plane curve. In this case the equations become

$$\frac{da}{ds} = f(s)b, \qquad \frac{db}{ds} = -f(s)a, \qquad \frac{dc}{ds} = 0.$$

Evidently, $a\,da + b\,db = 0$ and hence $a^2 + b^2 = k^2$, where k is an arbitrary constant. The equations may, then, be rewritten in the forms

$$\frac{da}{\sqrt{k^2 - a^2}} = f(s)ds, \quad \frac{db}{\sqrt{k^2 - b^2}} = -f(s)ds, \quad dc = 0,$$

and have, therefore, the general solution

$$a = k \sin(t + l), \quad b = k \cos(t + l), \quad c = m,$$

where

$$t = \int_0^s f(s)ds.$$

The particular solutions (52) which for $s = 0$ take on the prescribed values (53) are, in this case,

$$\begin{array}{lll} \alpha_1 = \cos t, & \beta_1 = -\sin t, & \gamma_1 = 0, \\ \alpha_2 = \sin t, & \beta_2 = \cos t, & \gamma_2 = 0, \\ \alpha_3 = 0, & \beta_3 = 0, & \gamma_3 = 1. \end{array}$$

They result from the general solution when the constants k, l, m are given in turn the values 1, $\pi/2$, 0; 1, 0, 0; 0, 0, 1.

It remains simply to substitute the values of α_1, α_2, α_3 in (55). Thus, we obtain the parametric representation

$$x_1 = \int_0^s \cos t\, ds, \quad x_2 = \int_0^s \sin t\, ds, \quad x_3 = 0$$

of the given plane curve in terms of the arc s and the curvature $1/R = f(s)$.

The integration of equations (51) in the general case depends on the solution of a so-called Riccati differential equation.*

EXERCISES

1. Show that the natural equation of the arch of the cycloid

$$x_1 = a(\theta - \sin\theta), \quad x_2 = a(1 - \cos\theta), \quad x_3 = 0, \quad 0 \leqq \theta \leqq 2\pi,$$

when the arc s is measured from the top of the arch, is $R^2 + s^2 = 16a^2$.

2. Find the natural equation of the involute of a circle:

$$x_1 = a(\cos\theta + \theta\sin\theta), \quad x_2 = a(\sin\theta - \theta\cos\theta), \quad x_3 = 0.$$

3. Integrate equations (51) in the case of a circular helix.

* See Eisenhart, *Differential Geometry*, p. 25.

22. Cylindrical helices. Consider in the (x_1, x_2)-plane a plane curve, D, restricted, if necessary, so that it has no points of inflection. The concave side of the curve is, then, always on one and the same side of the curve and the principal normal is always directed toward this side. Consequently, if we agree to trace the curve so that the concave side of it is always to our left, the binormal will be directed upward and have the direction cosines 0, 0, 1.

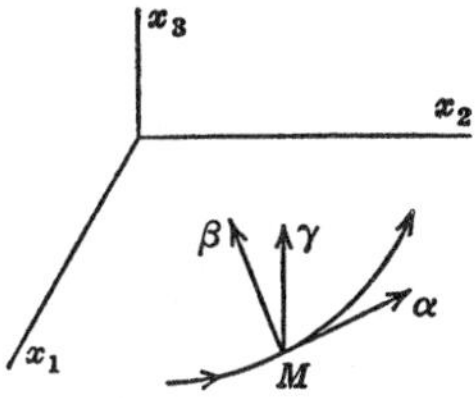

FIG. 7

The parametric equations of the curve D in terms of the arc σ, measured in the direction described, are of the form

$$D: \qquad x_1 = x_1(\sigma), \qquad x_2 = x_2(\sigma), \qquad x_3 = 0.$$

We find that the curvature, $1/r$, of D is

$$\frac{1}{r} = \sqrt{x_1''^2 + x_2''^2},$$

and that the vectors α, β, γ of the trihedral have the components x_1', x_2', 0; rx_1'', rx_2'', 0; 0, 0, 1, respectively.

Since $\gamma_3 = \alpha_1\beta_2 - \alpha_2\beta_1$, we have

$$(56) \qquad x_1'x_2'' - x_1''x_2' = \sqrt{x_1''^2 + x_2''^2}.$$

It is to be noted that, if we had traced D in the opposite direction, γ_3 would have been -1, and we should have had a minus sign before the right-hand member in (56).

Consider, now, the curve

$$(57) \qquad x_1 = x_1(\sigma), \quad x_2 = x_2(\sigma), \quad x_3 = \cot\theta\,\sigma, \quad 0 < \theta < \pi,$$

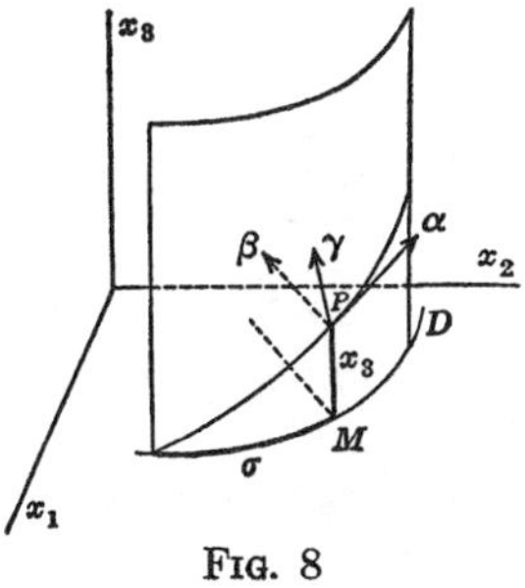

FIG. 8

where θ is a constant. This curve lies on the cylinder erected on the plane curve D as directrix and is the locus of a point P which moves on the cylinder so that its directed distance, x_3, from D is proportional to the arc σ of D; see Fig. 8. It is known as a (cylindrical) helix.

In writing the constant ratio of x_3 to σ in the form $\cot\theta$, we have placed no restriction on its value, inasmuch as θ may be any angle between 0 and π. In particular, this constant ratio, the so-called pitch of the helix, may be zero. Thus, we agree, for this paragraph, but not for later ones, to include among the helices the plane curves, other than straight lines.

From (57) we find that the components of x' and x'' are x_1', x_2', $\cot\theta$ and x_1'', x_2'', 0 respectively. Hence,

$$\begin{array}{lccc}
\widehat{x'\,x''}: & -\cot\theta\, x_2'', & \cot\theta\, x_1'', & x_1'x_2'' - x_1''x_2'; \\
\widehat{\widehat{x'\,x''}\,x'}: & \csc^2\theta\, x_1'', & \csc^2\theta\, x_2'', & 0 \quad ;
\end{array}$$

$$(x'\,|\,x') = \csc^2\theta, \qquad (\widehat{x'\,x''}\,|\,\widehat{x'\,x''}) = \csc^2\theta(x_1''^2 + x_2''^2),$$

where, in computing the components of $\widehat{\widehat{x'\,x''}\,x'}$, we have used the identities

$$x_1'^2 + x_2'^2 \equiv 1, \qquad x_1'x_1'' + x_2'x_2'' \equiv 0,$$

which follow from the fact that σ is the arc of D.

Thus, we obtain the results:

$$\begin{array}{llll}
 & \alpha: & \sin\theta\, x_1', & \sin\theta\, x_2', & \cos\theta\,; \\
 & \beta: & r\, x_1'', & r\, x_2'', & 0\,; \\
(58) & \gamma: & -r\cos\theta\, x_2'', & r\cos\theta\, x_1'', & \sin\theta\,;
\end{array}$$

$$\frac{1}{R} = \frac{\sin^2\theta}{r}, \qquad \frac{1}{T} = -\frac{\sin\theta\cos\theta}{r}, \qquad s = \csc\theta\,\sigma.$$

It is to be remarked that we have measured the arc s from the point $\sigma = 0$, that relation (56) is necessary to compute γ_3, and that the torsion has been obtained, without the direct use of x''', by means of the relation

$$\frac{d\beta_3}{ds} = -\frac{\alpha_3}{R} - \frac{\gamma_3}{T} \quad \text{or} \quad 0 = \frac{\cos\theta}{R} + \frac{\sin\theta}{T}.$$

As a first consequence of formulas (58), we note that, since $\alpha_3 = \cos\theta$, the helix cuts the rulings of the cylinder under the constant angle θ.

Suppose, conversely, that the directed tangents to a curve C, not a straight line, all make the same angle θ, $0 < \theta < \pi$, with a fixed directed line. Take this directed line as the positive axis of x_3, and choose the (x_1, x_2)-plane so that it contains at least one point P_0 of C. Project C on the (x_1, x_2)-plane, measure the arc σ of the projection, D, from the point P_0, and write the parametric equations of C in terms of σ: $x_1 = x_1(\sigma)$, $x_2 = x_2(\sigma)$, $x_3 = x_3(\sigma)$.

By hypothesis, $\alpha_3 = \cos\theta$ or $dx_3 = \cos\theta\, ds$, where s is the arc of C. Inasmuch as σ is the arc of D, $dx_1^2 + dx_2^2 = d\sigma^2$. Thus,

$$ds^2 = dx_1^2 + dx_2^2 + dx_3^2 = d\sigma^2 + \cos^2\theta\, ds^2.$$

Hence, if s is measured in the proper direction, $ds = \csc\theta\, d\sigma$. Then $dx_3 = \cot\theta\, d\sigma$ and $x_3 = \cot\theta\, \sigma$. The parametric representation of C is therefore of the form (57) and C is a helix.

Thus, we have established a first characteristic property of a helix:

THEOREM 1. *A curve, not a straight line, is a helix if and only if it makes a constant angle with a fixed line.*

Returning to formulas (58), we conclude, from $\beta_3 = 0$, that the principal normals of the helix (57) are all parallel to the (x_1, x_2)-plane. Conversely, if the principal normals of a curve are parallel to a plane, and we take the plane as the (x_1, x_2)-plane, then $\beta_3 = 0$. Hence, $d\alpha_3/ds = 0$, and $\alpha_3 = \text{const}$. The curve is, then, by Theorem 1, a helix.

THEOREM 2. *A necessary and sufficient condition that a curve be a helix is that its principal normals be parallel to a plane.*

From the formulas for $1/R$ and $1/T$ in (58), we find that the ratio of $1/T$ to $1/R$ is constant and equal to $-\cot\theta$.

Suppose, conversely, that, for a given curve, this ratio is constant, and write the constant in the form $-\cot\theta$, $0 < \theta < \pi$:

$$\frac{1}{T} + \frac{\cot\theta}{R} = 0. \tag{59}$$

We conclude, from the first and the third of the Frenet-Serret formulas, that $\cot\theta\, d\alpha + d\gamma = 0$. Hence

$$\cot\theta\,\alpha + \gamma = c, \tag{60}$$

where c is a triple of constants, not all zero. Taking the scalar product of each member of (60) with α, and then the scalar product of each member of (60) with itself, we find that $(\alpha|c) = \cot\theta$ and $(c|c) = \csc^2\theta$. Hence, $(\alpha|c)/\sqrt{c|c} = \cos\theta$; that is, the curve makes the constant angle θ with the vector c, and so is a helix.

THEOREM 3. *A curve is a helix if and only if the ratio of the torsion to the curvature is constant.*

The theorem says that equation (59) is the intrinsic equation of all helices.

EXERCISES

1. Show that the twisted cubic (4) is a helix if and only if its arc is a polynomial in t.

2. According to the previous exercise, the twisted cubic of § 10, Example 2, is a helix. Find the cylinder on which it is a helix and the angle under which it cuts the rulings of this cylinder.

3. Show that the curve $x = x(s)$ is a helix if and only if $(x_s'' \, x_s''' \, x_s^{\text{iv}}) = 0$. Suggestion. Prove that

$$-\frac{1}{R^5}\frac{d}{ds}\left(\frac{R}{T}\right) = (x_s'' \, x_s''' \, x_s^{\text{iv}}).$$

4. Integrate equations (51) in the case of a helix.

23. Bertrand curves. These are the curves whose principal normals are the principal normals of a second curve.

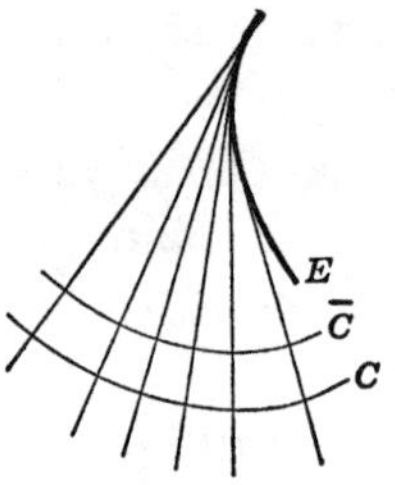

FIG. 9

Every plane curve has this property. For, if C is a plane curve and E is its evolute, that is, the envelope of its principal normals, then the involutes of E, that is, the curves cutting the tangent lines of E orthogonally, all have the same principal normals as C Moreover, the distance between corresponding points of two involutes, measured along the common normal, is constant.

Consider, now, an arbitrary curve C: $x = x(t)$, and assume that there exists a curve $\bar{C}$ which has the same principal normals as C. If $\bar{P}$: $(\bar{x}_1, \bar{x}_2, \bar{x}_3)$ is the point of $\bar{C}$ which corresponds to the arbitrary point P: (x_1, x_2, x_3) of C and if r is the directed distance from P to $\bar{P}$, measured along the principal normal to C at P, then the symbolic equation

$$\bar{x} = x + r\beta \tag{61}$$

constitutes a parametric representation of $\bar{C}$.

Evidently,

$$\bar{\alpha} = \frac{d\bar{x}}{d\bar{s}} = \frac{d\bar{x}}{ds}\frac{ds}{d\bar{s}} = \left(\alpha + r\frac{d\beta}{ds} + \frac{dr}{ds}\beta\right)\frac{ds}{d\bar{s}}. \tag{62}$$

By hypothesis, $(\bar{\alpha}\,|\,\beta) = 0$. Hence we find that $dr/ds = 0$. Thus r is a constant and (62) becomes

$$\bar{\alpha} = \left[\left(1 - \frac{r}{R}\right)\alpha - \frac{r}{T}\gamma\right]\frac{ds}{d\bar{s}}. \tag{63}$$

We proceed to distinguish two cases, according as $1/T \equiv 0$, or $1/T \not\equiv 0$.

C a plane curve. Equation (63) says in this case that $\bar{\alpha}$ is a multiple of α, and since both $\bar{\alpha}$ and α are unit vectors, the multiplier must be ± 1. Hence

$$\bar{\alpha} = \pm\,\alpha.$$

Then

$$\frac{\bar{\beta}}{\bar{R}} = \frac{d\bar{\alpha}}{d\bar{s}} = \pm\frac{d\alpha}{ds}\frac{ds}{d\bar{s}} = \pm\frac{\beta}{R}\frac{ds}{d\bar{s}},$$

whence $\bar{\beta} = \pm\,\beta$. Therefore, the curve $\bar{x} = x + r\beta$ has the same principal normals as the curve $x = x(t)$ for every constant value of r. (There is one exception in case the latter curve is a circle. What is it?)

C a twisted curve. In this case we have, from (63),

$$\frac{\bar{\beta}}{\bar{R}} = \left[\left(1 - \frac{r}{R}\right)\alpha - \frac{r}{T}\gamma\right]\frac{d^2s}{d\bar{s}^2} + \left[\frac{d}{ds}\left(1 - \frac{r}{R}\right)\alpha - \frac{d}{ds}\left(\frac{r}{T}\right)\gamma + \left(1 - \frac{r}{R}\right)\frac{\beta}{R} - \frac{r}{T}\frac{\beta}{T}\right]\left(\frac{ds}{d\bar{s}}\right)^2.$$

By hypothesis, $\bar{\beta} = \pm\,\beta$. Making this substitution and collecting the terms in α, β, γ, we obtain a relation of the form

$$A\alpha + B\beta + C\gamma = 0,$$

where A, B, C are scalar functions of t. If these functions were not all (identically) zero, the relation would say that the three vectors α, β, γ were parallel to a plane,—a contradiction. Hence, $A = 0$, $B = 0$, $C = 0$.

Writing out the relations $A = 0$ and $C = 0$, we have

$$\left(1 - \frac{r}{R}\right)\frac{d^2s}{d\bar{s}^2} + \frac{d}{ds}\left(1 - \frac{r}{R}\right)\left(\frac{ds}{d\bar{s}}\right)^2 = 0,$$

$$\frac{r}{T}\frac{d^2s}{d\bar{s}^2} + \frac{d}{ds}\left(\frac{r}{T}\right)\left(\frac{ds}{d\bar{s}}\right)^2 = 0.$$

Eliminating $ds/d\bar{s}$ and $d^2s/d\bar{s}^2$, the former of which is certainly not zero, we obtain the equation

$$\frac{r}{T}\frac{d}{ds}\left(1 - \frac{r}{R}\right) - \left(1 - \frac{r}{R}\right)\frac{d}{ds}\left(\frac{r}{T}\right) = 0,$$

or

$$\frac{d}{ds}\frac{1 - \frac{r}{R}}{\frac{r}{T}} = 0.$$

Integrating and writing the constant of integration in the form $\cot\theta$, we obtain, finally, the relation

$$\frac{1}{R} + \frac{\cot\theta}{T} = \frac{1}{r}. \tag{64}$$

Thus, a necessary condition that the twisted curve C be a Bertrand curve is that constants r and θ, $0 < \theta < \pi$, exist so that equation (64) is satisfied. This condition is sufficient: if (64) holds, the curve $\bar{C}$ defined by (61), where r is the constant in (64), has the same principal normals as the curve C. For, by use of (64), (63) can be rewritten in the form

$$\bar{\alpha} = (\cos\theta\,\alpha - \sin\theta\,\gamma)\frac{ds}{d\bar{s}}\frac{r}{T}\csc\theta.$$

Hence the vectors $\bar{\alpha}$ and $\cos\theta\,\alpha - \sin\theta\,\gamma$ are parallel. But the latter, as well as the former, is a unit vector. Hence

$$\bar{\alpha} = \pm(\cos\theta\,\alpha - \sin\theta\,\gamma), \tag{65}$$

and therefore

$$\frac{\bar{\beta}}{\bar{R}} = \frac{d\bar{\alpha}}{d\bar{s}} = \pm\left(\frac{\cos\theta}{R} - \frac{\sin\theta}{T}\right)\frac{ds}{d\bar{s}}\beta.$$

Thus, $\bar{\beta} = \pm\beta$, and the proposition is established.

Theorem 1. *A twisted curve is a Bertrand curve if and only if its curvature and torsion are connected by a linear relation of the form* (64) *in which r and $\cot\theta$ are constants.*

The distance between corresponding points P and $\bar{P}$ of a Bertrand curve C and its mate $\bar{C}$ is constant and equal to $|r|$. Moreover, the angle between the directed tangents at P and $\bar{P}$ is constant and equal either to θ or $\pi - \theta$. For, it follows from (65) that $(\bar{\alpha}\,|\,\alpha) = \pm\cos\theta$.

Suppose that the Bertrand curve C has a second mate $\bar{C}'$, that is, that $1/R$ and $1/T$ satisfy a second relation of the form (64), namely,

$$\frac{1}{R} + \frac{\cot\theta'}{T} = \frac{1}{r'}.$$

Solving this equation and (64), we find that $1/R$ and $1/T$ are constant. But then infinitely many relations of the form (64) exist; for, r may be given a value at pleasure, whereupon a value for $\cot\theta$ is determined. Thus:

Corollary. *A circular helix is the only twisted Bertrand curve which has more than one mate, and it has infinitely many mates.*

When the natural equation (64) of the twisted Bertrand curves is compared with that of the helices, it becomes evident that the only helices which are Bertrand curves are the circular helices.

Twisted curves of constant curvature. If $1/R$ is constant, equation (64) is satisfied by $r = R$ and $\theta = \pi/2$. Hence, a twisted curve C of constant curvature is a Bertrand curve; the curve $\bar{C}: \bar{x} = x + R\beta$, with the same principal normals as C, is the locus of the center of curvature of C; and the tangents to $\bar{C}$ and C at corresponding points are mutually perpendicular.

Since C is a Bertrand curve with $\bar{C}$ as a mate, $\bar{C}$ is a Bertrand curve with C as a mate. Moreover, $\bar{C}$ is a twisted curve as well as C, inasmuch as a mate of a plane curve is never twisted. Hence, there exists for $\bar{C}$ a relation of the form $1/\bar{R} + \cot\bar{\theta}/\bar{T} = 1/\bar{r}$, where $|\bar{r}|$ is the distance between corresponding points of $\bar{C}$ and its mate C and $\bar{\theta}$ is the angle between the tangents to $\bar{C}$ and C at these points. But we know that $|\bar{r}| = R$ and $\bar{\theta} = \pi/2$. Consequently, $1/\bar{R} = 1/R$ and so $\bar{r} = R$. In other words, $\bar{C}$ is a twisted curve of constant curvature (equal to that of C) and C is the locus of the center of curvature of $\bar{C}$.

We may summarize these results as follows:

Theorem 2. *The twisted curves of constant curvature may be paired as Bertrand mates. The curves of a pair have the same curvature, each is the locus of the center of curvature of the other, and the tangents to them at corresponding points are mutually perpendicular.*

Since the tangents at corresponding points are mutually perpendicular and the principal normals are the same, the normal plane to the one curve at a given point is the osculating plane of the other at the corresponding point.

EXERCISES

1. Establish the following symmetric relations between a Bertrand curve C and its mate $\bar{C}$:

$$\bar{\alpha} = \cos\theta\,\alpha - \sin\theta\,\gamma, \quad \bar{\beta} = -\beta, \quad \bar{\gamma} = -\sin\theta\,\alpha - \cos\theta\,\gamma,$$

$$\alpha = \cos\theta\,\bar{\alpha} - \sin\theta\,^{-}, \quad \beta = -\bar{\beta}, \quad \gamma = -\sin\theta\,\bar{\alpha} - \cos\theta\,\bar{\gamma},$$

$$\bar{x} = x + r\beta, \qquad x = \bar{x} + r\bar{\beta},$$

$$\frac{1}{\bar{R}} + \frac{\cot\theta}{\bar{T}} = \frac{1}{r}, \qquad \frac{1}{T}\frac{1}{\bar{T}} = \frac{\sin^2\theta}{r^2}.$$

Note that the choice of sign in (65) specifies a positive direction on $\bar{C}$ and that the assumption $\bar{\beta} = -\beta$, together with the application to $\bar{C}$ of the Frenet-Serret formulas in the standard form (43), may involve sacrificing, for $\bar{C}$, the convention that the curvature be non-negative.

2. Determine all the Bertrand curves each of which has a congruent curve as a mate.

3. Determine the curves whose binormals are the binormals of a second curve.

4. Find the natural equation of the curves whose principal normals are the binormals of a second curve.

EXERCISES ON CHAPTER II

1. Show that the tangents to a space curve and the locus of its center of curvature at corresponding points are mutually perpendicular.

2. Show that the principal normal of a twisted curve at a point P is tangent to the locus of the center of curvature when and only when $1/T = 0$ at P and is perpendicular to this locus when and only when $d(1/R)/ds = 0$ at P.

3. Prove that, if the twisted curve $x = x(s)$ has constant torsion, the curve

$$y = T\beta + \int \gamma\, ds$$

has constant curvature.

4. Show that, if the curve $x = x(s)$ is a helix, so also is the curve

$$y = R\alpha - \int \beta\, ds.$$

5. Show that, if the points of two curves are in one-to-one correspondence so that the tangents at corresponding points are parallel, the principal normals at corresponding points are parallel. Is the converse true?

6. The points of two curves are in one-to-one correspondence. Prove that, if the tangents at corresponding points are parallel, so also are the osculating planes, and conversely.

7. The points of two curves are in one-to-one correspondence so that the tangents at corresponding points are parallel. One of the curves has constant curvature, $1/a$, and the other constant torsion, $1/a$. Show that the locus of the midpoint of the line-segment joining corresponding points of the two curves is a Bertrand curve.

8. If two curves are reflections of one another in a point, or in a plane, their curvatures at corresponding points are equal and their torsions are negatives of one another.

9. If all the tangents to a curve go through a point, the curve is a straight line.

10. If all the osculating planes of a curve go through a point, the curve is a plane curve.

11. The *tangent indicatrix*, the *principal normal indicatrix*, and the *binormal indicatrix* of the curve $C: x = x(s)$ are the loci represented respectively by the symbolic equations

$$x = \alpha, \qquad x = \beta, \qquad x = \gamma.$$

Show that all three indicatrices lie on the sphere of unit radius whose center is at the origin, and that, if C is not a straight line, the first two are always curves, and the third is a curve unless C lies in a plane.

12. Show that the tangents to the tangent and binormal indicatrices at the points corresponding to a given point P of the twisted curve C are parallel to the principal normal to C at P and that, if both indicatrices are directed in the positive sense of this normal, the derivatives of their directed arcs, s_α and s_γ, with respect to the arc s of C are the curvature and torsion of C: $ds_\alpha/ds = 1/R$, $ds_\gamma/ds = 1/T$.

13. Show that a twisted curve is a helix if and only if the tangent indicatrix, or the binormal indicatrix, is a plane curve, and hence a circle.

14. Prove that the osculating circle at a point P has contact of at least the third order with the curve if and only if $1/T$ and $d(1/R)/ds$ vanish at P.

15. Investigate the nature of the locus of the center of curvature of a twisted curve in the neighborhood of a point at which $1/T$ and $d(1/R)/ds$ vanish; see Ex. 2.

CHAPTER III

CURVES AND SURFACES ASSOCIATED WITH A SPACE CURVE

24. Tangent surface of a space curve. The surface which is generated by the tangent lines to a curve C, not a straight line, is known as the tangent surface of C. If $x = x(s)$ is the symbolic representation of C in terms of its arc, the tangent surface of C is represented symbolically by the equation

$$y = x(s) + r\alpha(s), \tag{1}$$

where r is a parameter as well as s. For, when s is fixed and r varies, the point (y_1, y_2, y_3) defined by the equation traces the tangent line to C at the point whose coordinate is s. If s now varies, this point traces C and the tangent line, moving with it, sweeps out the tangent surface.

If C is a plane curve, the surface is simply a portion of the plane of C.

When C is a twisted curve, it is worth while to study the shape of the surface, particularly in the neighborhood of C. For this purpose, we choose arbitrarily a regular point P_0 of C, measure the arc s from P_0, and introduce the directed edges of the trihedral at P_0 as the coordinate axes. The coordinates (x_1, x_2, x_3) of an arbitrary point P of C in the neighborhood of P_0 are, then, given by the power series (46) of Chapter II, and the direction cosines $\alpha_1, \alpha_2, \alpha_3$ of the tangent at P are obtained by differentiating these power series term by term. Hence, the portion of the tangent surface generated by the tangents to C at the points in the neighborhood of P_0 is represented parametrically by the equations

$$\begin{aligned} y_1 &= \left(s - \frac{1}{6}\frac{1}{R_0^2}s^3 + \cdots\right) + \left(1 - \frac{1}{2}\frac{1}{R_0^2}s^2 + \cdots\right)r, \\ y_2 &= \left(\frac{1}{2R_0}s^2 + \frac{1}{6}\frac{d}{ds}\left(\frac{1}{R}\right)_0 s^3 + \cdots\right) \\ &\qquad + \left(\frac{1}{R_0}s + \frac{1}{2}\frac{d}{ds}\left(\frac{1}{R}\right)_0 s^2 + \cdots\right)r, \end{aligned} \tag{2}$$

$$y_3 = \left(-\frac{1}{6}\frac{1}{R_0T_0}s^3 + \cdots\right) + \left(-\frac{1}{2}\frac{1}{R_0T_0}s^2 + \cdots\right)r.$$

We look at the section of the surface by the normal plane of C at P_0, that is, by the plane $x_1 = 0$. A parametric representation of this section may be obtained from (2) by solving the equation $y_1 = 0$ for r and substituting the value thus found, namely,

$$r = -\frac{s - \dfrac{1}{6R_0^2}s^3 + \cdots}{1 - \dfrac{1}{2R_0^2}s^2 + \cdots} = -s - \frac{1}{3R_0^2}s^3 + \cdots,$$

in the expressions for y_2 and y_3. The representation turns out to be

$$y_1 = 0, \quad y_2 = -\frac{1}{2R_0}s^2 + \cdots, \quad y_3 = \frac{1}{3R_0T_0}s^3 + \cdots.$$

Neglecting the terms of the series which are not explicitly given, we conclude that the section, in the neighborhood of P_0, is approximately a semi-cubical parabola whose cusp is at P_0 and whose cuspidal tangent is the principal normal to C at P_0.

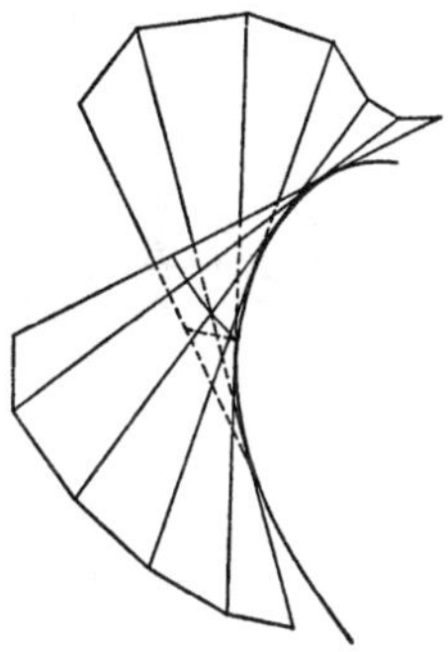

FIG. 10

It follows that the surface consists of two sheets which come together to form a sharp edge along the given curve. For this reason, the curve is called the *edge of regression* of the surface.

Since the two sheets of the surface are separated by the curve and $r = 0$ for the points of the curve, the values of r for the points on the one sheet are positive and those for the points on the other sheet are negative. Consequently, the one sheet is generated by the positive half-tangents of the curve, and the other by the negative half-tangents.

EXERCISES

1. Show that the tangent surface of a twisted cubic is a quartic surface.

Suggestion. Show that the tangent surface of $x_1 = at$, $x_2 = bt^2$, $x_3 = ct^3$ may be represented parametrically by the equations

$$y_1 = a(t + u), \quad y_2 = b(t^2 + 2ut), \quad y_3 = c(t^3 + 3ut^2),$$

and that elimination of the parameters t and u from these equations yields an equation of the fourth degree in y_1, y_2, y_3.

2. The osculating plane of a twisted curve C at a point P meets the tangent surface of C in the tangent line to C at P and a curve through P. Show that the curvature of the curve at P is three-quarters of the curvature of C at P.

25. Parametric representation of a surface. The equations

$$y_1 = x_1 + r\alpha_1, \qquad y_2 = x_2 + r\alpha_2, \qquad y_3 = x_3 + r\alpha_3$$

constitute a representation, in terms of the parameters r and s, of the tangent surface of the curve $x = x(s)$.

Similarly, the equations

$$(3) \quad x_1 = a \sin \phi \cos \theta, \qquad x_2 = a \sin \phi \sin \theta, \qquad x_3 = a \cos \phi$$

represent parametrically the sphere whose center is at the origin and whose radius is a. The parameters ϕ and θ are, respectively, the colatitude and longitude of a point on the sphere.

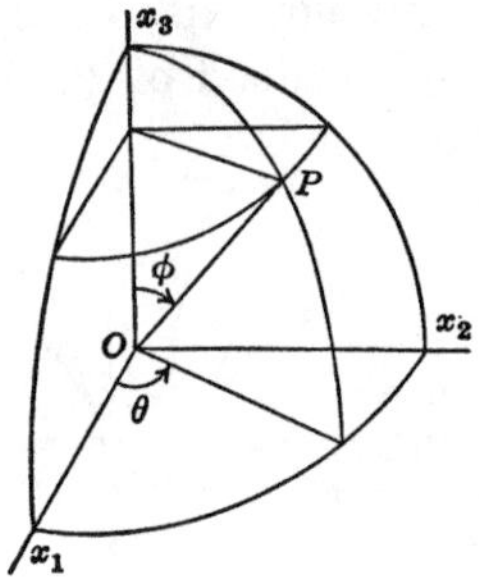

Fig. 11

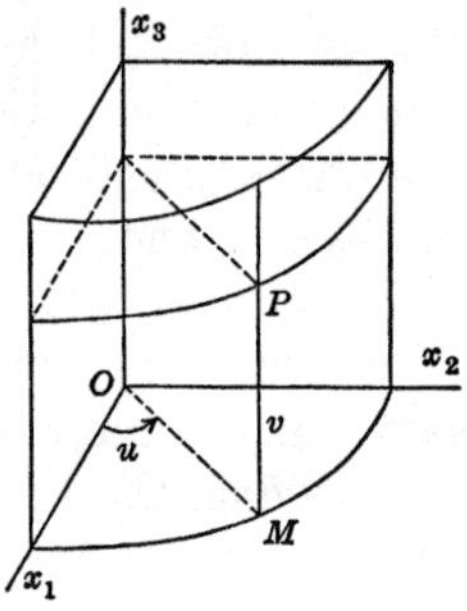

Fig. 12

Again, the circular cylinder erected on the circle in the (x_1, x_2)-plane with center at the origin and radius a, as directrix, has the parametric representation

$$(4) \qquad x_1 = a \cos u, \qquad x_2 = a \sin u, \qquad x_3 = v.$$

The equations

$$(5) \qquad x_1 = x_1(u, v), \qquad x_2 = x_2(u, v), \qquad x_3 = x_3(u, v),$$

where $x_1(u, v)$, $x_2(u, v)$, $x_3(u, v)$ are real, single-valued, analytic functions of the real variables u, v, with a common domain of definition, represent a surface, provided merely, as we shall show later, that $x_u \widehat{\ } x_v \not\equiv 0$, where x_u and x_v are the vectors $\partial x_1/\partial u$, $\partial x_2/\partial u$, $\partial x_3/\partial u$, and $\partial x_1/\partial v$, $\partial x_2/\partial v$, $\partial x_3/\partial v$.

The equation in x_1, x_2, x_3 of the surface (5) may be obtained by eliminating u, v from (5). This equation may, however, yield more than the parametric representation. For example, the elimination of ϕ, θ from equations (3) gives always the equation, $x_1^2 + x_2^2 + x_3^2 = a^2$, of the entire sphere, even when ϕ and θ are restricted so that equations (3) represent only a portion of the sphere.

Parametric curves. When $v = v_0$, that is, when v is set equal to an arbitrarily chosen constant, equations (5) represent, in general, a curve, which lies, of course, on the surface in question. Similarly, $u = u_0$ defines a curve on the surface. There are clearly infinitely many curves of each kind. They are called the parametric curves on the surface, and are designated as the curves v = constant and u = constant.

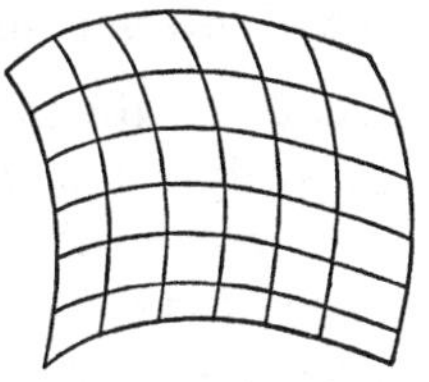

FIG. 13

The curves v = const. on the cylinder (4) are the circular cross-sections, whereas the curves u = const. are the rulings. In the case of the sphere (3), the curves ϕ = const. are the parallels of latitude and the curves θ = const. are the half-meridians.

Curvilinear coordinates. Since $x_1(u, v)$, $x_2(u, v)$, $x_3(u, v)$ are single-valued functions, there corresponds to a chosen pair of values u_0, v_0 of u, v a unique point on the surface (5). The converse is, however, not necessarily true. For example, to a given point of the cylinder (4) there corresponds an infinite number of pairs of values of u, v, of the form $u_0 + 2k\pi$, v_0, where $k = 0, \pm 1, \pm 2, \cdots$. We may, however, reduce the number to one by restricting u to lie in the interval $0 \leqq u < 2\pi$. Similarly, in the case of the arbitrary surface (5), we may, and do, restrict the domain of definition of the functions $x_1(u, v)$, $x_2(u, v)$, $x_3(u, v)$ so that to each point of the surface there corresponds, in general, a single pair of values of u, v.

In the case of the sphere (3), we restrict ϕ and θ to lie respectively in the intervals $0 \leqq \phi \leqq \pi$ and $0 \leqq \theta < 2\pi$. Then, to each point of the sphere corresponds a unique pair of values of ϕ, θ, in general. The north and south poles are exceptions, since θ is not determined for either of them. In fact, the equations $\phi = 0$ and $\phi = \pi$ represent, not curves, but simply these exceptional points.

Returning to our arbitrary surface (5), we now have, in general, a one-to-one correspondence between the points of the surface and the pairs of values of u, v. In other words, u, v constitute coordinates for a point on the surface. Since the coordinate lines $u =$ const. and $v =$ const. are generally curved, these coordinates, (u, v), are known as *curvilinear coordinates* on the surface.

Through a point (u_0, v_0) there pass, in general, just two parametric curves, one from each family, namely, the curve $u = u_0$ and the curve $v = v_0$, and the point may be thought of as the point of intersection of these two curves.

Curves on the surface. Just as a curve in the (x, y)-plane is defined by a functional relation between the coordinates (x, y), so a curve on the surface (5) is fixed by a functional relation between the coordinates (u, v). This relation may be expressed by means of an equation in u, v, such as $v = f(u)$ or $F(u, v) = 0$, or it may be represented parametrically: $u = u(t)$, $v = v(t)$. All the functions involved are, of course, to be analytic.

For example, since the parameter u in (4) is proportional to the arc of the directrix circle, measured from the point (1, 0, 0), the equation $v = ku$, $k \neq 0$, represents a circular helix on the cylinder.

Tangent plane at a point. By a tangent line to a surface at a point is meant the tangent line at the point to a curve on the surface which goes through the point.

We proceed to show that the tangent lines to the surface (5) at a point P for which $x_u \overset{\frown}{\ } x_v \neq 0$ lie in a plane. Let P be the point (u_0, v_0), and consider first the parametric curves which pass through it. The curve $v = v_0$ is represented parametrically, in terms of u, by $x = x(u, v_0)$. Hence the tangent to it at an arbitrary point has the direction components $x_u(u, v_0)$; in particular, the tangent at P has the direction components $x_u(u_0, v_0)$. Similarly, the tangent at P to the curve $u = u_0$ has the direction components $x_v(u_0, v_0)$.

The assumption that $x_u \overset{\frown}{\ } x_v \neq 0$ at P: (u_0, v_0) implies that the vectors x_u and x_v at P are proper vectors with distinct directions. We have just seen that these directions are those of the parametric curves. Consequently, the tangents at P to the parametric curves are distinct and determine a plane.

Consider, now, an arbitrary curve C passing through P and represent it, as a curve on the surface, by parametric equations, $u = u(t)$, $v = v(t)$, where $u_0 = u(t_0)$, $v_0 = v(t_0)$. As a curve in space, C has the representation $x = x(u(t), v(t))$. Hence, the tangent to C at an arbitrary point has the direction components dx/dt. But

$$\frac{dx}{dt} = \frac{\partial x}{\partial u}\frac{du}{dt} + \frac{\partial x}{\partial v}\frac{dv}{dt}.$$

Thus the tangent to C at P has the direction components

$$\left(\frac{dx}{dt}\right)_0 = x_u(u_0, v_0)\left(\frac{du}{dt}\right)_0 + x_v(u_0, v_0)\left(\frac{dv}{dt}\right)_0.$$

This equation says that the vector $(dx/dt)_0$, tangential to C at P, is a linear combination of the vectors x_u and x_v tangential to the parametric curves at P. Hence, the tangent to C at P lies in the plane determined by the tangents at P to the parametric curves.

We have thus proved that the tangent lines to the surface (5) at a point P for which $x_u \frown x_v \neq 0$ lie in a plane. This plane is called the *tangent plane* at P and the line perpendicular to it at P is known as the *normal* to the surface at P.

Since the normal at P is perpendicular to the tangents at P to the parametric curves, it has the direction components $x_u \frown x_v$. In other words, *the vector* $x_u \frown x_v$ *at* P *is a vector normal to the surface at* P.

Application to the tangent surface of a curve. For the surface (1): $y = x + r\alpha$, we have $y_r = \alpha$, $y_s = \alpha + (r/R)\beta$. Hence $y_r \frown y_s = (r/R)\gamma$. Thus, the normal to the surface at the point (r, s), $r \neq 0$, is parallel to the binormal of the curve at the point with the coordinate s. It follows that, for s fixed, that is, *for all the points of a ruling, the tangent plane is the same and is precisely the osculating plane of the curve at the point in which the ruling is tangent to the curve.*

The tangent planes to the surface are, then, the osculating planes of the curve. But any surface, not a plane, may be considered as the envelope of its tangent planes. Hence, *the tangent surface of a twisted curve is the envelope of the osculating planes of the curve.*

Ordinarily, a surface has a different tangent plane at each point and is therefore the envelope of a two-parameter family of planes. A surface which, like the tangent surface of a twisted curve, has only a one-parameter family of distinct tangent planes and so is the envelope of a one-parameter family of planes, is known as a *developable surface*.

26. Envelopes. Preparatory to a careful treatment of the envelope of a one-parameter family of planes, we shall give a descriptive, rather than rigorous, account of the envelope of a one-parameter family of surfaces. Let the equation of the family be

$$f(x_1, x_2, x_3, t) = 0, \tag{6}$$

where t is the parameter and $f(x_1, x_2, x_3, t)$ is a real, single-valued, analytic function of the four real variables.

We fix our attention on an arbitrary, but fixed, surface, S, of the family and assume that it is intersected by a neighboring surface, S', of the family in a curve C'. If S and S' correspond to the values t and $t + \Delta t$ of the parameter, the curve C' is represented by the simultaneous equations

$$f(x_1, x_2, x_3, t) = 0, \qquad f(x_1, x_2, x_3, t + \Delta t) = 0.$$

It is equally well represented by the equations

$$f(x_1, x_2, x_3, t) = 0, \qquad \frac{f(x_1, x_2, x_3, t + \Delta t) - f(x_1, x_2, x_3, t)}{\Delta t} = 0, \tag{7}$$

since the surface $f(x_1, x_2, x_3, t + \Delta t) - f(x_1, x_2, x_3, t) = 0$ goes through the curve common to the two surfaces $f(x_1, x_2, x_3, t) = 0$ and $f(x_1, x_2, x_3, t + \Delta t) = 0$ and meets neither in a further point.

When S' approaches S as a limit, that is, when $\Delta t \to 0$, the curve C' will, in general, approach a curve C. Since the limit of the left-hand member of the second equation in (7), when $\Delta t \to 0$, is $f_t(x_1, x_2, x_3, t)$, this curve C has the equations

$$f(x_1, x_2, x_3, t) = 0, \qquad f_t(x_1, x_2, x_3, t) = 0. \tag{8}$$

Inasmuch as the curve C' lies always on S, the curve C lies on S. It is known as the *characteristic curve* on S.

When the parameter t varies, the characteristic curve C will, in general, vary with S and generate a surface E. This surface E

is defined to be the *envelope* of the given family (6). It can, in fact, be proved that E is tangent to each surface S of the family at every point of the characteristic curve C on S.

Equations (8), when t is fixed, represent the characteristic curve on the corresponding surface of the family. The same equations, with t variable, represent the envelope E. The result of eliminating t from them is the equation, in x_1, x_2, x_3, of E.

Let us now consider the envelope of the characteristic curves, thought of as curves on the surface E. The characteristic C, given by (8), will, in general, have in common with the neighboring characteristic,

$$f(x_1, x_2, x_3, t + \Delta t) = 0, \qquad f_t(x_1, x_2, x_3, t + \Delta t) = 0,$$

one or more points P'. These points P' are precisely the points in which C is met by the curve

$$\begin{aligned} f(x_1, x_2, x_3, t + \Delta t) - f(x_1, x_2, x_3, t) &= 0, \\ f_t(x_1, x_2, x_3, t + \Delta t) - f_t(x_1, x_2, x_3, t) &= 0. \end{aligned}$$

Hence the points P which are the limits of the points P', when $\Delta t \to 0$, are the points in which C is met by the curve

$$f_t(x_1, x_2, x_3, t) = 0, \qquad f_{tt}(x_1, x_2, x_3, t) = 0,$$

or, what is the same thing, the points which the three surfaces

$$(9) \quad f(x_1, x_2, x_3, t) = 0, \quad f_t(x_1, x_2, x_3, t) = 0, \quad f_{tt}(x_1, x_2, x_3, t) = 0$$

have in common.

When the parameter t varies so that C runs through all the characteristic curves, the points P will, in general, vary and generate a curve R on the surface E. This curve R is defined as the envelope of the characteristics and is known as the *edge of regression* of the surface E. As a matter of fact, it can be shown that R is tangent to the characteristic C at each of the points P.

Equations (9), when t is fixed, define the points P on the characteristic C. Hence, when t is variable, these equations represent the edge of regression, R, and the result of solving them for x_1, x_2, x_3 is a parametric representation of R.

Example. It is evident geometrically that the envelope of the one-parameter family of spheres traced by a sphere of fixed radius

a whose center traverses a given curve $C: x = x(s)$ is the surface generated by a circle of radius a moving so that its center traces C and its plane is always normal to C.

To establish this result analytically, it suffices to show that the characteristic curve of the sphere with center at an arbitrary point P of C, namely

$$f \equiv (X - x | X - x) - a^2 = 0, \tag{10}$$

is the circle in which the sphere is cut by the normal plane to C at P. But the characteristic curve is the curve of intersection of the sphere and the surface

$$f_s \equiv -2 (X - x | \alpha) = 0, \tag{11}$$

and this surface is precisely the normal plane to C at P.

To obtain a parametric representation of the edge of regression of the envelope, we have to solve equations (10), (11), and

$$f_{ss} \equiv -2\left(X - x \middle| \frac{\beta}{R}\right) + 2 = 0 \tag{12}$$

for X_1, X_2, X_3. Equation (11) says that the vector $X - x$ at P lies in the normal plane to C at P and hence is a linear combination of β and γ: $X - x = B\beta + C\gamma$. Substituting this value of $X - x$ in (12) and (10), we find that $B = R$ and $C = \pm \sqrt{a^2 - R^2}$. Hence, the edge of regression of the envelope has the parametric representation

$$X = x + R\beta \pm \sqrt{a^2 - R^2}\,\gamma. \tag{13}$$

It is clear that, if R is always less than a, the edge of regression has two branches. On the other hand, if R is always greater than a, equation (13) fails to define real points.

If the given curve C is a twisted curve of constant curvature, $1/R$, and $a = R$, the edge of regression is the locus $\overline{C}: \overline{x} = x + R\beta$ of the center of curvature of C. Since the normal planes of C are the osculating planes of $\overline{C}$ (§ 23), it follows that the characteristic circles are the osculating circles of $\overline{C}$ and that the given spheres are the spheres having these circles as great circles. We shall learn shortly that these spheres are the so-called osculating spheres of $\overline{C}$; see § 29, Ex. 1. Hence it is not surprising that the edge of regression of their envelope is the curve $\overline{C}$.

EXERCISES

1. Find the envelope of the family of spheres which have as diameters the chords of an ellipse which are parallel to an axis of the ellipse.

2. From a point on a circle chords are drawn. Find the envelope of the spheres constructed on these chords as diameters. What can you say about the edge of regression?

27. Developable surfaces. In § 25, we showed that the tangent surface of a twisted curve is the envelope of a one-parameter family of planes. We shall now prove, conversely, that the envelope of a one-parameter family of planes is, in general, the tangent surface of a twisted curve.

Since the characteristic in a given plane of the family is the limiting position of the straight line in which this plane is met by a neighboring plane of the family, the characteristic curves are straight lines and the envelope E of the family of planes is a ruled surface.

The characteristic lines on E in general envelope a curve R and are, then, precisely the tangents to R. Hence, E is the tangent surface of the curve R.

What are the exceptions? In the first place, the given planes may all be parallel or they may all go through a fixed line. In the one case, there is no envelope E, and, in the other, E is a line. We exclude these cases, once and for all. The envelope E then exists and is a surface. The rulings of this surface will have a curve R as an envelope, unless they are all parallel or all go through a point. In the first case, E is a cylinder, and, in the second, E is a cone. Hence, we arrive at the final conclusion:

THEOREM 1. *A developable surface is the tangent surface of a twisted curve, or a cone, or a cylinder.*

It is a familiar fact that every cone or cylinder can be obtained by rolling a sheet of paper in the proper form. It is also a fact, which we shall prove later, that a sheet of a prescribed tangent surface can be obtained by drawing a suitably chosen plane curve on a piece of paper, cutting away the concave side, and rolling the portion which remains properly.* It follows that *every developable surface is*, as we say, *applicable to a plane* or can be developed on a plane, that is, can be rolled out on a plane without tearing or stretching.

* If the paper is creased lightly along each of a set of *half*-tangents to the curve, it will roll of itself into a sheet of a tangent surface.

Critical treatment. Let the family of planes be represented by the equation

$$f \equiv (a\,|\,x) + a_0 \equiv a_1(t)x_1 + a_2(t)x_2 + a_3(t)x_3 + a_0(t) = 0,$$

where $a_i(t)$, $i = 0, 1, 2, 3$, are analytic functions of t.

The characteristic line in the plane $f = 0$ is the line of intersection of the planes $f = 0$ and $f_t = 0$, namely, the line

$$(x\,|\,a) + a_0 = 0, \qquad (x\,|\,a') + a_0' = 0. \tag{14}$$

If $\widehat{a\,a'}$ were identically zero, it would follow, by § 13, Theorem 2, that the vector a is fixed in direction and hence that the planes of the family are all parallel. But this is contrary to hypothesis. Hence equations (14) actually define a characteristic line, whose direction is given by $\widehat{a\,a'}$.

The envelope, R, of the characteristic lines is obtained by solving the equations $f = 0$, $f_t = 0$, $f_{tt} = 0$, namely,

$$(x\,|\,a) + a_0 = 0, \quad (x\,|\,a') + a_0' = 0, \quad (x\,|\,a'') + a_0'' = 0, \tag{15}$$

for x_1, x_2, x_3. If $(a\,a'\,a'') \not\equiv 0$, the solution is possible by Cramer's rule, and the envelope exists. The envelope may, however, be only a point, in the case in which the functions of t obtained by the solution are all constants; then the rulings of the surface E all go through the point and E is a cone.

If $(a\,a'\,a'') \equiv 0$, the vector a is always parallel to a plane (§ 18, Theorem 3), and hence the planes of the family are all parallel to a line. The characteristic lines are, then, all parallel to this line and, since, by hypothesis, they are not all the same line, they generate a cylinder.

We proceed, finally, to justify the definitions of E and R. We prove, first, that the tangents to R, provided R is a curve, are actually the characteristic lines. If $x = x(t)$ is the parametric representation of R obtained by solving equations (15), then

$$(x(t)\,|\,a) + a_0 \equiv 0, \quad (x(t)\,|\,a') + a_0' \equiv 0, \quad (x(t)\,|\,a'') + a_0'' \equiv 0.$$

Differentiating the first and second of these identities with respect to t, and simplifying the results by means of the second and third, we find

$$(x'\,|\,a) \equiv 0, \qquad (x'\,|\,a') \equiv 0.$$

or $x' \equiv k\,\widehat{a\,a'}$. Hence the tangent to R at a point P has the direction of the characteristic line which goes through P and so actually is this characteristic line.

It remains to prove that the tangent planes to the surface E are the planes of the family. Mark a specific point on the characteristic line C and visualize the curve which it traces when C generates E. If $z = z(t)$ is the parametric representation of this curve in terms of the parameter t, the symbolic equation

$$y = z + r\,\widehat{a\,a'},$$

where both r and t vary, represents the surface E.

The plane of the family through the characteristic line C will be the tangent plane to E all along C if we can show that the normals to E in the points of C are all parallel to the vector a, that is, if we can show that $(a\,|\,y_r) \equiv 0$ and $(a\,|\,y_t) \equiv 0$. Now $y_r = \widehat{a\,a'}$ and $y_t = z' + r\,\widehat{a\,a''}$. Evidently, $(a\,|\,y_r) \equiv 0$; and $(a\,|\,y_t) \equiv 0$ if $(a\,|\,z') \equiv 0$.

Since the point z moves always as a point of C, its coordinates always satisfy both of the equations (14). Therefore

$$(a\,|\,z) + a_0 \equiv 0, \qquad (a'\,|\,z) + a_0' \equiv 0.$$

Differentiating the first of these identities and using the second to simplify the result, we find that $(a\,|\,z') \equiv 0$. Hence, the proof is complete.

Example. In the case of the family of planes

$$f \equiv t^3 - 3\,t^2x_1 + 3\,tx_2 - x_3 = 0,$$

we have

$$f_t \equiv 3\,t^2 - 6\,tx_1 + 3\,x_2 = 0, \qquad f_{tt} \equiv 6\,t - 6\,x_1 = 0.$$

Solving the three equations for x_1, x_2, x_3, we find, as the edge of regression, the twisted cubic $x_1 = t$, $x_2 = t^2$, $x_3 = t^3$. Hence, the envelope of the planes is the tangent surface of this twisted cubic.

EXERCISES

1. Find the envelope and edge of regression of the one-parameter family of planes

$$\sin t\,x_1 - \cos t\,x_2 + \tan\theta\,x_3 - at = 0.$$

2. Find the edge of regression of the developable surface which is tangent to the hyperbolic paraboloid $x_1^2 - x_2^2 = ax_3$ along the curve in which the paraboloid is cut by the circular cylinder $x_1^2 + x_2^2 = a^2$.

28. Rectifying developable. There are three important families of planes associated with a space curve, namely, the families of osculating, rectifying, and normal planes. The osculating planes envelope the tangent surface, whose edge of regression is the curve itself. In discussing the envelopes of the rectifying and normal planes, we shall find the following lemma useful.

LEMMA. *In the theory of the family of surfaces* $f(x_1, x_2, x_3, t) = 0$, *the equations* $f = 0$, $f_t = 0$, $f_{tt} = 0$ *may be replaced by the equations* $f = 0$, $\phi f_t = 0$, $(\phi f_t)_t = 0$, *where* ϕ *is any function of* t, *not zero.*

For, $(\phi f_t)_t \equiv \phi f_{tt} + \phi' f_t = 0$ reduces, since $f_t = 0$ and $\phi \neq 0$, to $f_{tt} = 0$.

In the case of the rectifying planes of the curve $x = x(s)$, we have

$$f \equiv (X - x|\beta) = 0, \tag{16}$$

$$-Rf_s \equiv \left(X - x\,\middle|\,\alpha + \frac{R}{T}\gamma\right) = 0, \tag{17}$$

$$-(Rf_s)_s \equiv \left(X - x\,\middle|\left(\frac{1}{R} + \frac{R}{T^2}\right)\beta + \frac{d}{ds}\left(\frac{R}{T}\right)\gamma\right) - 1 = 0. \tag{18}$$

From equations (16) and (17) it is evident that the characteristic line of the rectifying plane goes through the point x and has the direction of the vector $\gamma - (R/T)\alpha$. It has, therefore, the parametric representation

$$X = x + r\left(\gamma - \frac{R}{T}\alpha\right). \tag{19}$$

To find the point of the characteristic line which lies on the edge of regression, we determine r so that X, as given by (19), satisfies (18). We find that $rd(R/T)/ds = 1$, and hence obtain, as the coordinates of the point in question,

$$X = x + \frac{\gamma - \dfrac{R}{T}\alpha}{\dfrac{d}{ds}\left(\dfrac{R}{T}\right)}. \tag{20}$$

When s varies as well as r, the symbolic equation (19) represents the envelope of the rectifying planes: the *rectifying devel-*

opable of the given curve. This developable clearly passes through the curve.

Equation (20), when s is variable, represents the edge of regression of the rectifying developable. This equation has, however, no meaning when $R/T = \text{const}$. But, then, the given curve is a helix and the rectifying developable is the cylinder on which the helix lies; for, the principal normal to the helix at a point is the normal to the cylinder at the point (§ 22), and hence the rectifying plane is tangent to the cylinder.

If this cylinder is rolled out on a plane, the helix, since it meets the rulings of the cylinder under a constant angle, becomes a straight line in the plane. This property of the helix holds for every curve. *The process of rolling the rectifying developable of a curve out on a plane straightens out the curve into a straight line,* or, in other words, rectifies the curve. The proof of the fact in the general case we shall give later. Furthermore, we shall also establish the converse: If a curve on a developable surface, other than a straight line, is rectified by rolling the surface out on a plane, the developable is the rectifying developable of the curve.

EXERCISE

Show that $R/T = as + b$, $a \neq 0$, is the natural equation of the curves whose rectifying developables are cones, that is, of the curves on cones which are rectified when the cones are rolled out on planes.

29. Polar developable. Osculating sphere. We have still to consider the envelope of the normal planes of a curve $x = x(s)$, not a straight line. Here, we have

$$f \equiv (X - x \,|\, \alpha) = 0, \tag{21}$$

$$Rf_s \equiv (X - x \,|\, \beta) - R = 0, \tag{22}$$

$$(Rf_s)_s \equiv \left(X - x \,\middle|\, -\frac{\alpha}{R} - \frac{\gamma}{T}\right) - \frac{dR}{ds} = 0. \tag{23}$$

It is clear from equations (21) and (22) that the characteristic line of the plane $f = 0$ goes through the point $x + R\beta$ and has the direction $\widehat{\alpha\,\beta}$, or γ. In other words, the characteristic line of the normal plane goes through the center of curvature, C, and is parallel to the binormal. It is known as the *polar line,* and the surface which it generates, the envelope of the normal planes, is called the *polar developable.*

The symbolic equation

$$X = x + R\beta + r\gamma \tag{24}$$

represents the polar line or the polar developable, according as we think simply of r, or of both r and s, as variable.

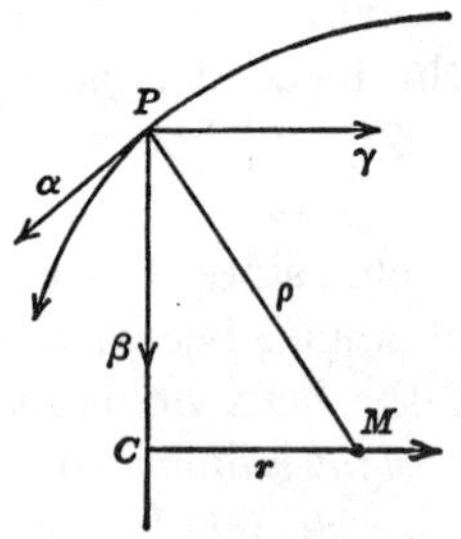

FIG. 14

The result of substituting X, from (24), into (23) is $r/T + dR/ds = 0$. Consequently, if the given curve is twisted, the polar developable has an edge of regression, defined by

$$X = x + R\beta - T\frac{dR}{ds}\gamma, \tag{25}$$

and hence is never a cylinder. On the other hand, the polar developable of a plane curve is always a cylinder; the normal planes are all perpendicular to the plane of the curve and envelope the cylinder whose directrix is the locus of the center of curvature of the curve.

Osculating sphere. We proceed to find, if it exists, the sphere through the osculating circle at a regular point, P, which has contact of the highest order with the curve.

The center M: (X_1, X_2, X_3) of an arbitrary sphere through this circle is on the polar line and hence has the coordinates (24), where r is the directed distance CM. If P': (y_1, y_2, y_3) is a point of the curve neighboring to P: (x_1, x_2, x_3), the distance, D, from P' to the sphere is

$$D = \pm (MP' - \rho),$$

where ρ is the radius of the sphere. Hence

$$\pm (MP' + \rho)D = \overline{MP'}^2 - \rho^2,$$

where

$$\overline{MP'}^2 = (X - y | X - y), \qquad \rho^2 = R^2 + r^2.$$

By Chapter II, (45),

$$y = x + \alpha\sigma + \frac{\beta}{R}\frac{\sigma^2}{2!} + \left(\frac{d}{ds}\left(\frac{1}{R}\right)\beta - \frac{\alpha}{R^2} - \frac{\gamma}{RT}\right)\frac{\sigma^3}{3!} + \cdots.$$

Hence

$$X - y = R\beta + r\gamma - \alpha\sigma \\ - \frac{\beta}{R}\frac{\sigma^2}{2!} - \left(\frac{d}{ds}\left(\frac{1}{R}\right)\beta - \frac{\alpha}{R^2} - \frac{\gamma}{RT}\right)\frac{\sigma^3}{3!} + \cdots$$

and

$$(X - y \mid X - y) = R^2 + r^2 - \left(R\frac{d}{ds}\left(\frac{1}{R}\right) - \frac{r}{RT}\right)\frac{\sigma^3}{3} + \cdots.$$

Thus

$$\pm (MP' + \rho)D = \frac{1}{R}\left(\frac{dR}{ds} + \frac{r}{T}\right)\frac{\sigma^3}{3} + \cdots. \tag{26}$$

If $1/T$ is not zero at the point P, it follows that there is a unique sphere which has contact of at least the third order with the curve at P, namely, the sphere for which $r = -T(dR/ds)$. The center of this sphere is in the point (25) and its radius is given by

$$\rho^2 = R^2 + \left(T\frac{dR}{ds}\right)^2. \tag{27}$$

It is called the *osculating sphere* of the curve at the point P.

THEOREM 1. *The center of the osculating sphere of a twisted curve at a regular point P at which $1/T \neq 0$ is the point on the edge of regression of the polar developable which corresponds to P. The locus of the center of the osculating sphere is the edge of regression of the polar developable.*

If $1/T = 0$ and $dR/ds \neq 0$ at P, the coefficient of σ^3 in (26) is never zero, and all the spheres in question have contact of the second order with the curve. On the other hand, we know that the osculating plane, which may be thought of as a degenerate sphere, has contact of at least the third order with the curve; see § 19.

If $1/T$ and dR/ds both vanish at P, the coefficient of σ^3 in (26) is always zero. In this case, all the spheres under discussion, including the osculating plane, have contact of at least the third order with the curve.

We define the osculating sphere, S, at a point P of a twisted curve at which $1/T = 0$ as the limit of the osculating sphere, S', at a neighboring point P', when P' approaches P as a limit. Evidently, S is either a sphere or a plane, never a point. Moreover, S must have contact of at least the third order with the curve, since this is true always of S'. It follows, then, from the foregoing considerations, that S is the osculating plane at P if $dR/ds \neq 0$ at P and may be the osculating plane or a sphere if $dR/ds = 0$ at P.

The osculating sphere of a plane curve ($1/T \equiv 0$), that is, the sphere or plane which has contact of highest order with the curve, is always the plane of the curve, unless the curve is a circle ($dR/ds \equiv 0$). The osculating sphere of a circle is undefined.

Spherical curves. The polar developable of a twisted curve is a cone when and only when its edge of regression (25) is a point. But it is readily shown that $dX/ds \equiv 0$ if and only if

$$R + T\frac{d}{ds}\left(T\frac{dR}{ds}\right) \equiv 0. \tag{28}$$

From Theorem 1, it follows that the center of the osculating sphere is, then, always at the vertex of the cone and so is fixed. Moreover, the radius of the osculating sphere is constant; for, it is readily shown that $d\rho^2/ds \equiv 0$ when (28) holds. Hence, the osculating sphere is one and the same sphere for all points of the curve, and the curve lies on this sphere. Conversely, if a twisted curve lies on a sphere, the center of the osculating sphere is fixed, and relation (28) follows. Thus (28) is the natural equation of the twisted curves which lie on spheres, or, as they are called, the spherical curves.

EXERCISES

1. Show that a twisted curve of constant curvature is characterized by the property that the edge of regression of the polar developable coincides with the locus of the center of curvature.

2. Show that the osculating sphere of a twisted curve is always of the same size if and only if the curve is spherical or has constant curvature.

3. Find α, β, γ, $1/R$, $1/T$ for the edge of regression of the polar developable of a twisted nonspherical curve. Show that if the curve is a helix, the edge of regression is a helix, and conversely.

30. Involutes. An involute of the curve C: $x = x(s)$, not a straight line, is a curve which cuts each tangent to C at right angles. It lies, therefore, on the tangent surface,

$$y = x + r\alpha,$$

of C. An arbitrary curve on this surface, other than a ruling, has an equation of the form $r = r(s)$. Since a vector in the direction of the curve is

$$(29) \qquad \frac{dy}{ds} = \alpha + \frac{r}{R}\beta + \frac{dr}{ds}\alpha,$$

the curve cuts the tangents to C orthogonally if and only if $(dy/ds \,|\, \alpha) = 0$. But, this condition reduces to

$$dr + ds = 0 \quad \text{or} \quad r = k - s,$$

and hence we are led to the conclusion:

THEOREM 1. *The curve $x = x(s)$ has infinitely many involutes, namely,*

$$(30) \qquad y = x + (k - s)\alpha.$$

If I_1 and I_2 are two distinct involutes, represented respectively by the equations $r_1 = k_1 - s$ and $r_2 = k_2 - s$, then $r_2 - r_1 = k_2 - k_1$. In other words:

THEOREM 2. *The distance between corresponding points of two involutes is constant.*

In the equation $r + s = k$ of the involute I, s is the arc of the given curve, measured from the fixed point P_0 to the arbitrary point P, and r is the tangential distance from P to the corresponding point Q of I. Since the equation says that $r + s$ is constant and equal to k, it follows that, if a string of length k, with one end fastened at P_0 and originally coincident with the curve, is unwound so that it always remains on the tangent surface, the free end traces the involute I. By varying the length of the string, all the involutes of C may be obtained in this way.

When $dr/ds = -1$, we find, from (29), that dy/ds is a multiple of β. Thus, the tangent to the involute I at Q is parallel to the principal normal to C at P.

31. Evolutes. An evolute of the curve C: $x = x(s)$, not a straight line, is a curve whose tangents cut C orthogonally, that is, a curve which is the envelope of normals to C.

By a family of normals to C we shall mean a one-parameter family which consists of one normal at each point of C. If the normals of the family have an envelope, the surface which they generate is the tangent surface of this envelope. Thus the families of normals which envelope evolutes of C are included among the families of normals which generate developable surfaces or planes. We shall find these latter families of normals first, and then select from among them those which actually have envelopes.

An arbitrary unit vector η normal to C at the point $P: (x_1, x_2, x_3)$ may be written as a linear combination of the vectors β and γ at P. We have, namely, $\eta = \cos\phi\,\beta + \sin\phi\,\gamma$, where ϕ is the angle through which β must be rotated, in the direction toward γ, in order to coincide with η.

If ϕ is thought of as a single-valued, analytic function of the arc s of C, the symbolic equation

$$\eta = \cos\phi\,\beta + \sin\phi\,\gamma \tag{31}$$

represents a family of unit vectors normal to C and the lines on which these vectors lie constitute a family of normals to C.

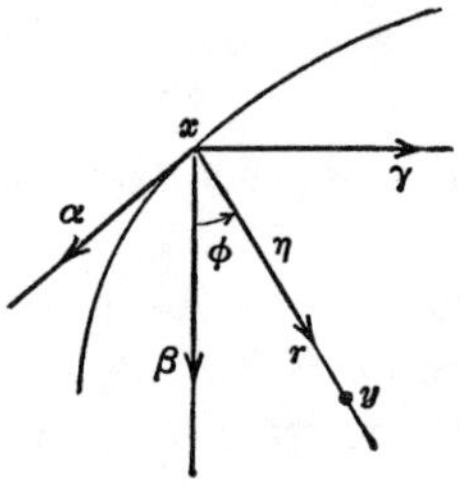

FIG. 15

THEOREM 1. *A necessary and sufficient condition that the family of normals to C defined by* (31) *generate a developable surface or a plane is that*

$$\frac{d\phi}{ds} - \frac{1}{T} = 0. \tag{32}$$

The surface generated by the normals of the family, expressed parametrically in terms of the arc s of C and the directed distance r along the rulings, has the symbolic representation

$$y = x + r\eta. \tag{33}$$

If the surface is a developable or a plane, the tangent plane is the same at all the points of each ruling. In other words, the direction of the vector $\overparen{y_s\, y_r}$ normal to the surface is independent of r. Since $y_s = \alpha + r\eta'$, where $\eta' = d\eta/ds$, and $y_r = \eta$,

$$\overparen{y_s\, y_r} = \overparen{\alpha\,\eta} + r\,\overparen{\eta'\,\eta}. \tag{34}$$

Evidently, this vector is, in direction, independent of r only if $\overparen{\eta'\,\eta} = k\,\overparen{\alpha\,\eta}$ or $\overparen{\eta' - k\alpha\;\eta} = 0$, that is, only if the scalar $k(s)$ exists so that the vector $\eta' - k\alpha$ is parallel to the vector η. But $\eta' - k\alpha$ is perpendicular to η; for, $(\alpha|\eta) = 0$ and, since η is a unit vector, $(\eta'|\eta) = 0$. Hence, $\eta' - k\alpha$ is a null vector and $\eta' = k\alpha$.

Differentiating η, as given by (31), we have

$$(35)\qquad \eta' = -\frac{\cos\phi}{R}\alpha + \left(\frac{d\phi}{ds} - \frac{1}{T}\right)(-\sin\phi\,\beta + \cos\phi\,\gamma).$$

Therefore, η' is a multiple of α only if relation (32) holds.

Conversely, if (32) holds, then η' is, by (35), a multiple of α, the vector $y_s\,y_r$ has a direction independent of r, and the surface is a developable or a plane.

From (32), we have

$$(36)\qquad \phi(s) = \int_0^s \frac{ds}{T} + c,$$

where c is an arbitrary constant. Thus, there are infinitely many values of $\phi(s)$, each two differing by a constant. Hence we conclude the following proposition.

THEOREM 2. *There are infinitely many families of normals to a given curve which generate developable surfaces or planes. The angle between corresponding normals of any two of these families is constant.*

We next determine when the developable is a cylinder or a cone. If it is a cylinder, η is always the same vector, that is, $\eta' \equiv 0$. We conclude, then, from (35) and (32), that $\phi \equiv \pi/2$ and $1/T \equiv 0$. Hence, C is a plane curve and the family of normals in question consists of the binormals.

If the rulings of the developable all go through a point, then $r = r(s)$ exists so that $y = x + r\eta$ represents a point. Making use of (32) and (35), we have

$$(37)\qquad \frac{dy}{ds} = \alpha + r\eta' + r'\eta = \left(1 - r\frac{\cos\phi}{R}\right)\alpha + r'\eta.$$

Hence, $dy/ds \equiv 0$ implies that $r' \equiv 0$ and r is a constant. This means that the point is at a constant distance from all the points of C. Therefore C lies on a sphere and the normals in question are normal to the sphere. What more can be said if, in particular, C is a circle?

COROLLARY. *If C is a plane curve, its binormals are all parallel. If C is a spherical curve, the normals to C which are also normals of a sphere on which C lies go through a point. Otherwise, the normals* *of a family of normals to C which generate a developable or a plane envelope a curve,—an evolute of C.*

From these facts we draw the following conclusion.

THEOREM 3. *Every curve, other than a circle, has infinitely many evolutes.*

The curve $y = x + r(s)\eta$ is an evolute of C if the tangential vector dy/ds has the direction of η. According to equation (37), which takes account of the fundamental condition (32), this is the case only if $r = R/\cos \phi$. Hence, an arbitrary evolute of C has the parametric representation

$$y = x + R \sec \phi \, \eta = x + R\beta + R \tan \phi \, \gamma, \tag{38}$$

where ϕ is given by (36).

Comparing (38) with (24), we conclude:

THEOREM 4. *The evolutes of a curve all lie on the polar developable of the curve.*

We return, finally, to Theorem 2, and restate its content in another way.

THEOREM 5. *If C is an orthogonal trajectory of a one-parameter family of lines which generates a developable or a plane, and each line is rotated through a constant angle about the tangent to C at the point in which it meets C, the lines in their new positions still generate a developable or a plane.*

EXERCISES

1. *The binormals of a curve never have an envelope.*

2. *The principal normals of a curve have an envelope only when the curve is a plane curve*, not a circle, and the envelope is, then, the locus of the center of curvature of the curve.

3. Show that the twisted evolutes of a plane curve are helices on the cylinder which is the polar developable of the curve.

4. Find α, β, γ, $1/R$, $1/T$ for an involute of a twisted curve C, and show that the polar developable of the involute is the rectifying developable of C. Find all the evolutes of the involute and verify the fact that one of them is C itself.

5. Prove that involutes of a twisted curve are plane curves if and only if the twisted curve is a helix.

CHAPTER IV

FUNDAMENTALS OF THE THEORY OF SURFACES

32. Parametric representation. The equations

$$(1) \qquad x_1 = x_1(u, v), \qquad x_2 = x_2(u, v), \qquad x_3 = x_3(u, v)$$

of § 25 do not always represent a surface. They may represent only a point, in case the three functions x_1, x_2, x_3 are all constants. Or, they may represent a curve, in case these functions are all expressible as functions of a single variable. For example, the manifold of points represented by

$$(2) \qquad x_1 = u + v, \qquad x_2 = (u + v)^2, \qquad x_3 = (u + v)^3$$

is precisely the same as that represented by $x_1 = t$, $x_2 = t^2$, $x_3 = t^3$, and so is a curve.

Each two of the functions in (2) are functions of one another; the second and the third are respectively the square and the cube of the first, and the cube of the second is equal to the square of the third.

If one of two functions is a function of the other, we say that the two functions are functionally dependent. It is a fact, the proof of which need not concern us, that *the functions* $f(u, v)$, $\phi(u, v)$ *are functionally dependent or both are constants if and only if the determinant* $f_u\phi_v - f_v\phi_u$, *known as their Jacobian, vanishes identically.*

Let the reader show that the Jacobian of each two of the functions in equations (2) is identically zero.

It is geometrically evident that the manifold of points represented by equations (1) is a curve or a point if and only if the projection of the manifold on each of the coordinate planes is a curve or a point. The projection of the manifold on a coordinate plane is represented by two of the equations, for example, by $x_1 = x_1(u, v)$, $x_2 = x_2(u, v)$, and is a curve or a point when and only when the two functions involved are functionally dependent or constants. But a necessary and sufficient condition that this be the case is that the Jacobian of the two functions vanish.

Hence, equations (1) represent a curve or a point if and only if the Jacobians of each two of the three functions involved are identically zero.

If not all of these Jacobians vanish identically, equations (1) represent a surface. Suppose, for example, that the Jacobian of the functions x_1, x_2 is not identically zero. This means that the coordinates x_1, x_2 of the variable point $(x_1, x_2, 0)$ are not functionally related or constants. Hence, the projection on the (x_1, x_2)-plane of the manifold represented by (1) is a region, and the manifold, itself, is a surface.

Since it may be readily verified that the Jacobians of the pairs of functions x_2, x_3, x_3, x_1, and x_1, x_2 in (1) are precisely the components of the triple $\overset{\frown}{x_u\, x_v}$, we may state our result as follows:

THEOREM 1. *Equations* (1) *represent a surface if and only if* $\overset{\frown}{x_u\, x_v} \neq 0$.

Even when equations (1) represent a surface, as we shall henceforth assume, there may be points on the surface for which $\overset{\frown}{x_u\, x_v} = 0$. In contrast to them, we shall call a point at which $\overset{\frown}{x_u\, x_v} \neq 0$ a *regular point* of the surface.

A point may fail to be regular on account of a singularity of the surface or on account of a singular behavior of the parametric representation.

For example, in the case of the tangent surface $y = x + r\alpha$ of a twisted curve, the points of the curve fail to be regular since the curve is a singular locus, an edge, of the surface. We have, in fact, $\overset{\frown}{y_r\, y_s} = (r/R)\gamma$, so that, when $r = 0$, $\overset{\frown}{y_r\, y_s} = 0$.

On the other hand, the north and south poles of the sphere $x_1 = a \sin\phi \cos\theta$, $x_2 = a \sin\phi \sin\theta$, $x_3 = a\cos\phi$, fail to be regular because they are singular points, not of the sphere, but of the parametric representation; see § 25. For the north pole, $\phi = 0$; and it is easily proved that, when $\phi = 0$, $x_\theta = 0$ and hence $\overset{\frown}{x_\phi\, x_\theta} = 0$.

It has been shown, in § 25, that, at a regular point P: (u, v) of the surface (1), the tangent lines really do lie in a plane and that the surface actually has a normal, whose direction is that of the vector $\overset{\frown}{x_u\, x_v}$. On the other hand, there is no guarantee that tangent plane and normal exist at a point which fails to be regular. At the vertex of a cone, for example, they certainly do not exist.

Change of parameters. The result of substituting for u and v in (1) the functions

$$u = u(\bar{u}, \bar{v}), \qquad v = v(\bar{u}, \bar{v}) \tag{3}$$

effects a change from the coordinates (u, v) on the surface to new coordinates $(\bar{u}, \bar{v})$. The functions $u(\bar{u}, \bar{v}), v(\bar{u}, \bar{v})$ are to be thought of as real, single-valued, analytic functions of the real variables $\bar{u}, \bar{v}$, defined throughout a certain domain of values for $(\bar{u}, \bar{v})$. Furthermore, in order to keep the correspondence between the pairs of coordinate values and the points of the surface in general one-to-one, it is assumed that equations (3) establish a one-to-one correspondence, in general, between the pairs of values of the old coordinates (u, v) and the permissible pairs of values of the new coordinates $(\bar{u}, \bar{v})$. This implies that the Jacobian of the two functions involved is not identically zero:

$$\Delta \equiv \begin{vmatrix} \dfrac{\partial u}{\partial \bar{u}} & \dfrac{\partial v}{\partial \bar{u}} \\ \dfrac{\partial u}{\partial \bar{v}} & \dfrac{\partial v}{\partial \bar{v}} \end{vmatrix} \not\equiv 0.$$

For, if Δ were identically zero, the pairs of values of (u, v) resulting from (3) would be restricted by a relation, $F(u, v) = 0$, which would permit them to yield at most only the points of a curve on the surface.

Since the new parametric representation $x = \bar{x}(\bar{u}, \bar{v})$ of the surface is obtained by substituting $u = u(\bar{u}, \bar{v})$, $v = v(\bar{u}, \bar{v})$ in $x = x(u, v)$, we have

$$\bar{x}(\bar{u}, \bar{v}) \equiv x(u, v), \quad \text{where} \quad u = u(\bar{u}, \bar{v}), v = v(\bar{u}, \bar{v}).$$

Hence,

$$\bar{x}_{\bar{u}} = x_u \frac{\partial u}{\partial \bar{u}} + x_v \frac{\partial v}{\partial \bar{u}}, \qquad \bar{x}_{\bar{v}} = x_u \frac{\partial u}{\partial \bar{v}} + x_v \frac{\partial v}{\partial \bar{v}},$$

and

$$\widehat{\bar{x}_{\bar{u}}\,\bar{x}_{\bar{v}}} = \widehat{x_u\,x_v} \frac{\partial u}{\partial \bar{u}} \frac{\partial v}{\partial \bar{v}} + \widehat{x_v\,x_u} \frac{\partial v}{\partial \bar{u}} \frac{\partial u}{\partial \bar{v}},$$

or

$$\widehat{\bar{x}_{\bar{u}}\,\bar{x}_{\bar{v}}} = \Delta\, \widehat{x_u\,x_v}. \tag{4}$$

This important relation tells us that, if $\widehat{x_u\, x_v} \neq 0$, then $\widehat{\bar{x}_{\bar{u}}\, \bar{x}_{\bar{v}}} \neq 0$, unless $\Delta = 0$. In other words, a point P which is regular with respect to the coordinates (u, v) is also regular with respect to the coordinates $(\bar{u}, \bar{v})$ unless $\Delta = 0$ at P.

Henceforth, we shall restrict ourselves to the regular points of a surface, unless the contrary is explicitly stated.

33. Linear element. First fundamental form. Let there be given on the surface S:

$$x = x(u, v), \tag{5}$$

a curve C: $u = u(t)$, $v = v(t)$. The differential of arc, ds, of C is given by

$$ds^2 = (dx\,|\,dx) = (x_u du + x_v dv\,|\,x_u du + x_v dv),$$

or

$$ds^2 = (x_u\,|\,x_u)du^2 + 2\,(x_u\,|\,x_v)du dv + (x_v\,|\,x_v)dv^2, \tag{6}$$

where $du = u'dt$, $dv = v'dt$. Thus, the arc s, measured from the point $t = t_0$ in the direction of increasing t, is

$$s = \int_{t_0}^{t} \sqrt{(x_u\,|\,x_u)u'^2 + 2\,(x_u\,|\,x_v)u'v' + (x_v\,|\,x_v)v'^2}\, dt. \tag{7}$$

The expression on the right-hand side of (6) is a quadratic form, that is, a homogeneous polynomial of degree two, in du, dv. It is known as the *first fundamental differential quadratic form* of S, or, since it defines the differential of arc for an arbitrary curve C on S, as the *linear element* of S.

It is customary to denote the coefficients in (6) by E, F, G. We then have

$$ds^2 = Edu^2 + 2\,Fdudv + Gdv^2, \tag{8}$$

where

$$E = (x_u\,|\,x_u), \qquad F = (x_u\,|\,x_v), \qquad G = (x_v\,|\,x_v). \tag{9}$$

The discriminant, $EG - F^2$, of the quadratic form we shall denote by D^2:

$$D^2 = EG - F^2.$$

By means of (9) and Lagrange's identity (§ 6), we find that

$$D^2 = (x_u\, x_v \,|\, x_u\, x_v).$$

Consequently, $D = \sqrt{EG - F^2}$ is positive at every regular point.

It does not follow from (8) that ds is always equal to the positive square root of the quadratic form, since ds would then be positive for both directions along the curve C. As a matter of fact, we have, from (7),

$$ds = \sqrt{E\left(\frac{du}{dt}\right)^2 + 2\,F\frac{du}{dt}\frac{dv}{dt} + G\left(\frac{dv}{dt}\right)^2}\,dt,$$

and hence

$$ds = \pm\,\sqrt{Edu^2 + 2\,Fdudv + Gdv^2},$$

where the plus or minus sign is to be taken according as $dt > 0$ or $dt < 0$, that is, according as du, dv are the differentials of u and v in the positive, or in the negative, direction along C.

The linear element enables us to find, not only the lengths of arcs of curves, but also angles between curves. In giving applications of the latter type, we begin with the parametric curves.

Parametric curves. In the Cartesian (x, y)-plane, we speak of the line $y = 0$ as the x-axis, inasmuch as x is the variable coordinate on it. For the same reason, we shall call the curves $v = \text{const.}$ on S, along which the coordinate u varies, the *u-curves*; and the curves $u = \text{const.}$, the *v-curves*.

The positive direction on a parametric curve shall be the direction in which the variable coordinate increases. It follows, from (8), that the differentials of arc of an arbitrary u-curve and an arbitrary v-curve are, respectively,

$$ds = \sqrt{E}\,du, \qquad ds = \sqrt{G}\,dv; \tag{10}$$

for, in the first case, $v = \text{const.}$ or $dv = 0$; and in the second, $du = 0$.

The directed tangent to a directed curve C: $u = u(s)$, $v = v(s)$ has the direction cosines

$$\frac{dx}{ds} = x_u\frac{du}{ds} + x_v\frac{dv}{ds}, \tag{11}$$

where s is the arc of C measured in the positive direction along C.

If C is, in particular, a u-curve, then $dv = 0$ and $du/ds = 1/\sqrt{E}$; and, if C is a v-curve, $du = 0$ and $dv/ds = 1/\sqrt{G}$. Hence, the unit vectors tangent at the point P: (u, v) to the parametric curves which pass through P are, respectively,

$$\frac{x_u}{\sqrt{E}}, \quad \frac{x_v}{\sqrt{G}}, \tag{12}$$

where x_u, x_v, E, G are evaluated for P.

It follows that, if the angle between the directed parametric curves at P is ω, $0 < \omega < \pi$, then

$$\cos \omega = \frac{F}{\sqrt{EG}}, \qquad \sin \omega = \frac{D}{\sqrt{EG}}. \tag{13}$$

Thus, the vanishing of F at P is the condition that these curves cut orthogonally. Hence, if $F \equiv 0$, each curve of the one family of parametric curves meets every curve of the other family orthogonally, and conversely. This result we state as follows:

THEOREM 1. *The parametric curves on the surface S form an orthogonal system if and only if $F = 0$.*

The convention introduced here by writing $F = 0$ instead of $F \equiv 0$ we shall adhere to henceforth. To indicate that a function $\phi(u, v)$ vanishes identically, we shall write merely "$\phi(u, v) = 0$." On the other hand, to express the fact that $\phi(u, v)$ vanishes at a particular point P, we shall write "$\phi(u, v) = 0$ at P."

Arbitrary curves. Let there be given, through the point P: (u, v), two directed curves, C and C'. The unit vectors tangent to them at P are

$$\alpha = x_u \frac{du}{ds} + x_v \frac{dv}{ds}, \qquad \alpha' = x_u \frac{\delta u}{\delta s} + x_v \frac{\delta v}{\delta s},$$

where du/ds, dv/ds pertain to C and $\delta u/\delta s$, $\delta v/\delta s$ pertain to C', and these derivatives, as well as x_u and x_v, are evaluated for P. Hence if θ, $0 \leqq \theta \leqq \pi$, is the angle between the curves, the values of $\cos \theta$ and $\sin \theta$, obtained by use of the formulas $\cos \theta = (\alpha \mid \alpha')$, $\sin \theta = \sqrt{\alpha\, \alpha' \mid \alpha\, \alpha'}$, are

$$\begin{aligned} \cos \theta &= E \frac{du}{ds}\frac{\delta u}{\delta s} + F\left(\frac{du}{ds}\frac{\delta v}{\delta s} + \frac{dv}{ds}\frac{\delta u}{\delta s}\right) + G \frac{dv}{ds}\frac{\delta v}{\delta s}, \\ \sin \theta &= D \left| \frac{du}{ds}\frac{\delta v}{\delta s} - \frac{dv}{ds}\frac{\delta u}{\delta s} \right|, \end{aligned} \tag{14}$$

where the bars denote the absolute value of the expression which they enclose.

If we take du, dv in the positive direction along C, and δu, δv in the positive direction along C', we may write, instead of (14),

$$(15)\qquad \begin{aligned} \cos\theta &= \frac{Edu\delta u + F(du\delta v + dv\delta u) + Gdv\delta v}{\sqrt{Edu^2 + 2Fdudv + Gdv^2}\sqrt{E\delta u^2 + 2F\delta u\delta v + G\delta v^2}}, \\ \sin\theta &= \frac{D\,|du\delta v - dv\delta u|}{\sqrt{Edu^2 + 2Fdudv + Gdv^2}\sqrt{E\delta u^2 + 2F\delta u\delta v + G\delta v^2}}. \end{aligned}$$

The first of these formulas implies the following result.

THEOREM 2. *A necessary and sufficient condition that the curves C and C' intersect orthogonally at the point P is that, at P,*

$$(16)\qquad Edu\delta u + F(du\delta v + dv\delta u) + Gdv\delta v = 0.$$

Area of a region on the surface. The area of a closed region on the surface is given by the double integral

$$(17)\qquad A = \iint Ddudv,$$

extended over the region of the (u, v)-plane whose points correspond to the points of the given region on the surface. In other words, $dA = Ddudv$ is the element of area of the surface referred to the curvilinear coordinates (u, v).

To establish these facts, we must show that the area, ΔA, of the curvilinear quadrilateral formed by the parametric lines $u = u$, $u = u + \Delta u$ and $v = v$, $v = v + \Delta v$ is equal to $D\Delta u\Delta v$, to within infinitesimals of higher order. The lengths of the sides of the quadrilateral which issue from the vertex P: (u, v) are, to within terms of higher order, $\sqrt{E}\,\Delta u$ and $\sqrt{G}\,\Delta v$, and the sine of the angle ω at P is $D/\sqrt{EG}$. Hence, if the quadrilateral were a rectilinear parallelogram, its area would be

$$(\sqrt{E}\,\Delta u)(\sqrt{G}\,\Delta v)(D/\sqrt{EG})$$

or $D\Delta u\Delta v$. Consequently, if we make the reasonable assumption that its actual area, ΔA, differs from the area of the parallelogram by an infinitesimal of higher order, the desired conclusion is obtained.

EXERCISES

1. Show that any two v-curves on the surface

$$x_1 = u \cos v, \quad x_2 = u \sin v, \quad x_3 = \log \cos u + v$$

cut equal segments from all the u-curves.

2. Find the first fundamental forms of the sphere and the cylinder of § 25, verifying in each case the fact that the parametric curves form an orthogonal system.

3. Show that the curve $\theta = \tan \alpha \log \tan (\phi/2) + \theta_0$ on the sphere goes through the point $(\theta_0, \pi/2)$ on the equator and cuts the meridians under the constant angle α, that is, is the general rhumb line or loxodrome on the sphere. Prove that its total length is finite and equal to $\pi a \sec \alpha$.

4. Show that, if C is the acute angle under which the great circle cut from the sphere by the plane $x_1 = x_3 \cot \alpha$, $0 < \alpha < \pi$, intersects the half-meridian $\theta = \theta_0$, then $\cos C = \sin \theta_0 \sin \alpha$.

5. Prove that the area of the triangle on the sphere bounded by the great circle of Ex. 4 and the half-meridians $\theta = \theta_1$ and $\theta = \theta_2$ is $a^2(A + B + C - \pi)$, where A, B, C are the angles of the triangle in radians. Consider first the case of the right triangle obtained by taking $\theta_1 = 0$.

34. Directions at a point. If the surface S is the (x_1, x_2)-plane, referred to u and v as rectangular coordinates, that is, if $x_1 = u$, $x_2 = v$, $x_3 = 0$, and C is an arbitrary curve, defined either by a parametric representation $u = u(t)$, $v = v(t)$ or by an equation $f(u, v) = 0$, then the direction of C at a point P is determined by the value of dv/du at P. For, the value of dv/du at P is precisely the slope of C at P.

The situation is similar at a point P on an arbitrary surface S. Let C be a curve, defined by $f(u, v) = 0$ or by $u = u(t)$, $v = v(t)$, which goes through P. According to (11), the tangent to C at P has the direction of the vector

$$x_u + \frac{dv}{du} x_v, \tag{18}$$

where x_u, x_v, dv/du are evaluated for P. This vector is a linear combination of the vectors x_u, x_v, tangent to the parametric curves at P, and just which direction it has in the tangent plane to S at P is fixed by dv/du. Thus, the direction of C at P is actually determined by the value of dv/du at P. In this case, however, dv/du cannot, in general, be interpreted as a slope.

In order to obtain a clearer idea of the situation, we digress a moment to discuss coordinates for the lines through a point P

in a plane. Of course, the slopes of the lines constitute as simple coordinates as can be imagined. But we are interested in coordinates of a more general type, which will help us to interpret dv/du in (18).

Choose two of the given lines as lines of reference, denoting them by L_0 and L_*, and giving to each a sense. Mark the point Q_0 on L_0 at a given positive directed distance a from P, draw through Q_0 the line M parallel to L_* and having the same sense, and mark the point Q_1 on M at the given positive directed distance b from Q_0. Next introduce, as the coordinate λ of a point Q on M, the ordinary Cartesian coordinate of Q referred to Q_0 as origin and to Q_0Q_1 as unit distance; in other words, take as λ the ratio of the directed distance $\overline{Q_0Q}$ to b: $\lambda = \overline{Q_0Q}/b$. Every point on M has, then, a unique coordinate λ, and every number is the coordinate of a unique point; in particular, Q_0 has the coordinate 0, and Q_1 the coordinate 1.

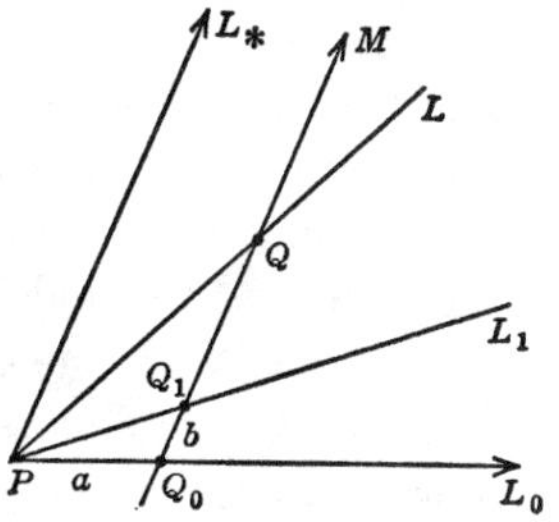

FIG. 16

Turning our attention to the lines through P, we take as the coordinate of a line L the coordinate, λ, of the point Q in which L meets M. Then every number is the coordinate of a specific line, and every line except L_* has a specific coordinate. When L approaches L_*, the coordinate, λ, of L becomes infinite.

If L_0 and L_* are mutually perpendicular and $a = b$, the coordinate λ of L is precisely the slope of L with respect to L_0 and L_* as rectangular axes. For, in this case, the slope of L is the ratio $\overline{Q_0Q}/a$, and, since $\overline{Q_0Q} = b\lambda$ and $b = a$, this ratio is equal to λ.

Suppose, now, that P is the point (x_1, x_2, x_3) on the surface S, that L_0, L_* are respectively the directed tangents to the u- and v-curves through P, and that a and b are taken as the values of $\sqrt{E}$ and $\sqrt{G}$ at P. The point Q on the line L with the coordinate λ may be reached from P by proceeding along L_0 to Q_0 and thence along M to Q. The unit vectors in the positive directions of L_0 and M are $x_u/\sqrt{E}$, $x_v/\sqrt{G}$, and $\overline{PQ_0} = \sqrt{E}$, $\overline{Q_0Q} = \sqrt{G}\,\lambda$. Hence, the coordinates of Q are given by (§ 7)

$$y = x + \sqrt{E}\frac{x_u}{\sqrt{E}} + \sqrt{G}\,\lambda\frac{x_v}{\sqrt{G}},$$

and the vector $y - x$ in the direction of the line L with coordinate λ is

$$x_u + \lambda x_v. \tag{19}$$

Comparison of this result with (18) furnishes a complete geometric interpretation of dv/du. The value of dv/du at P is precisely the coordinate, λ, of the tangent line to the curve C at P, as just defined. This value, then, actually determines the direction of C at P. We agree to speak of it as the coordinate of the direction or, simply, as the direction. For example, we might perfectly well talk of a curve through the point (1, 2) with the direction 3, and we shall speak frequently of a curve through the point P: (u, v) with the direction dv/du.

It is well to emphasize that we have introduced in the pencil of tangent lines at each point of the surface S a direction coordinate, and that the systems of reference for the direction coordinates at two distinct points may look entirely different. At one point, the direction coordinate may be a slope, whereas, at another, it may be of general type.

The direction coordinate at every point of S is a slope if and only if $F = 0$ and $E = G$. The parametric curves then form an orthogonal system of a special type which we shall discuss later; see § 47.

Angle between two directions. It is worth while now to review formulas (15) and (16). Equation (16) is essentially an equation in dv/du and $\delta v/\delta u$, since it can be rewritten in the form

$$E + F\left(\frac{dv}{du} + \frac{\delta v}{\delta u}\right) + G\frac{dv}{du}\frac{\delta v}{\delta u} = 0. \tag{20}$$

Hence it constitutes a necessary and sufficient condition that the two directions dv/du, $\delta v/\delta u$ at P be mutually perpendicular.

From (15), we conclude that the angles θ between the two directions at P with the direction coordinates λ_1, λ_2 are given by

$$\cos\theta = \pm\frac{E + F(\lambda_1 + \lambda_2) + G\lambda_1\lambda_2}{\sqrt{E + 2F\lambda_1 + G\lambda_1^2}\sqrt{E + 2F\lambda_2 + G\lambda_2^2}}. \tag{21}$$

There are, in this case, two angles θ, $0 \leqq \theta \leqq \pi$, inasmuch as a direction at P fixes merely a tangent line to S at P, without regard to sense.

35. Families and systems of curves. By a family of curves on the surface S: $x = x(u, v)$ is meant a one-parameter set of curves on S which can be represented by an equation of the form

$$f(u, v) = c, \tag{22}$$

where $f(u, v)$ is a single-valued analytic function and c is an arbitrary constant. The parametric curves $u =$ const. constitute, for example, a family of curves.

If (u_0, v_0) is a regular point of S and $f(u_0, v_0) = c_0$, the curve $f(u, v) = c_0$ goes through the point (u_0, v_0), and is evidently the only curve of the family (22) with this property. Hence, there is a unique curve of a family which passes through a given regular point.

Equivalent to (22) is the equation $f_u du + f_v dv = 0$, inasmuch as the result of integrating this equation is precisely (22).

Consider, now, the equation

$$M(u, v)du + N(u, v)dv = 0, \tag{23}$$

where $M(u, v)$, $N(u, v)$ are single-valued, analytic functions. Since there exists a function $I(u, v)$ such that $IMdu + INdv$ is the total differential of a function $f(u, v)$, this equation is also equivalent to an equation of the form (22) and so represents a family of curves on S.

Equation (23), since it may be written in the form $dv/du = - M/N$, determines at each point of S a direction, namely, the direction of the curve of the family which goes through the point. Conversely, if there is associated with each point P of S a direction in such a way that this direction varies in an analytic manner when P traces S, that is, so that the direction is expressible in the form $dv/du = - M/N$, where $M(u, v)$, $N(u, v)$ are analytic functions, then there exists a unique family of curves which have the given directions as their tangent directions, namely, the family of curves defined by the differential equation $Mdu + Ndv = 0$. We may, therefore, speak of this family as the family of curves in the direction dv/du or $- M/N$.

LEMMA. *The two families of curves $f(u, v) =$ const. and $\phi(u, v) =$ const. are identical if and only if $f(u, v)$, $\phi(u, v)$ are functionally dependent.*

For, the directions $dv/du = -f_u/f_v$ and $\delta v/\delta u = -\phi_u/\phi_v$ are the same at each point P when and only when $f_u/f_v = \phi_u/\phi_v$ or $f_u\phi_v - f_v\phi_u = 0$.

Systems of curves. Two distinct families of curves,

$$(24)\quad f(u, v) = \text{const.},\quad \phi(u, v) = \text{const.},\quad f_u\phi_v - f_v\phi_u \neq 0,$$

constitute a system of curves. The parametric curves, $u = \text{const.}$ and $v = \text{const.}$, for example, form a system.

There are two curves of a system through each regular point, one from each family. If these curves are always perpendicular, every curve of the one family cuts every curve of the other family at right angles, and the system is called an *orthogonal system.*

THEOREM 1. *The system of curves consisting of the families of curves in the directions dv/du, $\delta v/\delta u$ is orthogonal if and only if*

$$(25)\qquad Edu\delta u + F(du\delta v + dv\delta u) + Gdv\delta v = 0.$$

The theorem follows from (16). The identity is an identity in u, v; for, $E, F, G, dv/du$, and $\delta v/\delta u$ are all functions of u, v.

If the system consists of the families (24), then $dv/du = -f_u/f_v$, $\delta v/\delta u = -\phi_u/\phi_v$ and condition (25) becomes

$$(26)\qquad Ef_v\phi_v - F(f_v\phi_u + f_u\phi_v) + Gf_u\phi_u = 0.$$

The equations $M_1du + N_1dv = 0$, $M_2du + N_2dv = 0$ represent a system of curves provided $M_1N_2 - M_2N_1 \neq 0$. This system is equally well represented by the single equation

$$(M_1du + N_1dv)(M_2du + N_2dv) = 0,$$

the left-hand side of which is a quadratic form in du, dv. Conversely, every equation of the type

$$(27)\qquad l(u, v)du^2 + 2\,m(u, v)dudv + n(u, v)dv^2 = 0,$$

whose left-hand member can be factored into two real, non-proportional, linear forms, represents a system of curves. It is assumed, of course, that l, m, n are analytic functions of u and v.

THEOREM 2. *The system* (27) *is orthogonal if and only if*

$$(28)\qquad En - 2\,Fm + Gl = 0.$$

For, the directions dv/du, $\delta v/\delta u$ of the two families constituting the system (27) are the roots of the quadratic equation in dv/du obtained by dividing (27) by du^2. Hence,

$$\frac{dv}{du} + \frac{\delta v}{\delta u} = -2\frac{m}{n}, \qquad \frac{dv}{du}\frac{\delta v}{\delta u} = \frac{l}{n},$$

and substitution of these expressions in (20) yields the desired condition.

Orthogonal trajectories of a family. If a family of curves is given, there always exists a second family whose curves cut those of the given one orthogonally. It is known as the family of orthogonal trajectories of the given family.

THEOREM 3. *The family of orthogonal trajectories of the family of curves* $M du + N dv = 0$ *is represented by*

$$(EN - FM)du + (FN - GM)dv = 0. \tag{29}$$

For, the directions $\delta v/\delta u$ and dv/du of the given and required families must satisfy (25), and (25) becomes (29) when $\delta v/\delta u = -M/N$.

If the given family were represented by a finite equation, $f(u, v) = \text{const.}$, we should replace M and N in (29) by f_u and f_v respectively.

Example. The involutes of the space curve $x = x(s)$, since they are the orthogonal trajectories of the rulings of the tangent surface of the curve (§ 30), may be determined by means of equation (29).

For the tangent surface $y = x + r\alpha$, $y_r = \alpha$, $y_s = \alpha + (r/R)\beta$, and hence $E = 1$, $F = 1$, $G = 1 + (r/R)^2$. Since the rulings of the surface are the curves $s = \text{const.}$, they are represented by the equation $M dr + N ds = 0$, where $M = 0$, $N = 1$. Consequently, by (29), the orthogonal trajectories of the rulings have the differential equation $dr + ds = 0$ and hence the finite equation $r + s = \text{const.}$

EXERCISES

1. The surface

$$x_1 = u + a_1 v, \quad x_2 = u^2 + a_2 v, \quad x_3 = u^3 + a_3 v, \qquad (a|a) = 1,$$

is the cylinder obtained by drawing through the points of the twisted cubic $x_1 = u$, $x_2 = u^2$, $x_3 = u^3$ lines with the direction cosines a_1, a_2, a_3. Find the finite equation of the orthogonal trajectories of the rulings. Hence show that the cylinder is a cubic cylinder except in the case $a_1 = a_2 = 0$.

2. The surface

$$x_1 = u \cos v, \quad x_2 = u \sin v, \quad x_3 = a \cosh^{-1}\frac{u}{a}$$

is the *catenoid*, the surface of revolution generated by the rotation of the catenary about its axis. Find the finite equation of the loxodromes on it, that is, the curves making constant angles with the catenaries.

3. Show that, for every choice of the function $m(u, v)$, the differential equation $du^2 + 2\,mdudv + (a^2 - u^2)dv^2 = 0$ represents on the catenoid an orthogonal system of curves.

4. Show that a twisted curve which cuts the rulings of a cone of revolution under a constant angle is a helix on a cylinder whose directrix curve is a logarithmic spiral.

5. Find the differential equation of the system of curves which bisect the angles between the parametric curves on the surface $x = x(u, v)$.

36. The directed normal. Second fundamental form. Since the components of $\widehat{x_u\, x_v}$ are direction components of the normal to S at P: (u, v), and since $(\widehat{x_u\, x_v} \,|\, \widehat{x_u\, x_v}) = D^2$, direction cosines of the normal are given by $\widehat{x_u\, x_v}/D$ or $-\widehat{x_u\, x_v}/D$, according to the sense in which the normal is directed. We agree to choose that sense which corresponds to the direction cosines $\widehat{x_u\, x_v}/D$ and to denote these direction cosines by $\zeta_1, \zeta_2, \zeta_3$. Then

$$\zeta = \frac{\widehat{x_u\, x_v}}{D} \tag{30}$$

is a unit vector at P which has the same direction and sense as the directed normal. We shall call it *the normal vector* at P.

Taking the inner product of each side of (30) with ζ, we find

$$D = (x_u\, x_v\, \zeta). \tag{31}$$

Consequently,

$$\left(\frac{x_u}{\sqrt{E}}\, \frac{x_v}{\sqrt{G}}\, \zeta\right) > 0.$$

This inequality says that the normal to S has been so directed that the tangent vector to the u-curve, the tangent vector to the v-curve, and the normal vector have, in the order given, the same disposition as the coordinate axes; or, what is the same thing, that the direction of rotation from the tangent vector to the u-curve to the tangent vector to the v-curve, as viewed from the terminal point of the normal vector, is counterclockwise. To establish this fact, it suffices to show that, if the unit vectors α: $(1, 0, 0)$, β: $(\beta_1, \beta_2, 0)$, γ: $(0, 0, 1)$, at the origin are given, then the direction of rotation from the positive axis of x_1 to β, as viewed from a point on the positive half of the x_3-axis, is counterclockwise or clockwise, according as $(\alpha\,\beta\,\gamma) > 0$ or < 0. It is evident that the direction of rotation is counterclockwise or clockwise according as the sine of the directed angle ϕ in Fig. 17 is positive or negative. But $(\alpha\,\beta\,\gamma) = \beta_2$ and $\beta_2 = \sin\phi$. Hence, the contention is proved.

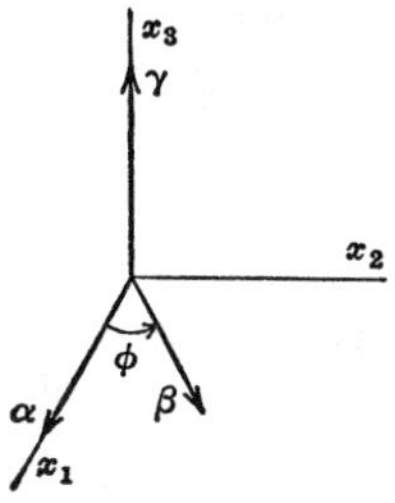

FIG. 17

Some important identities. Differentiating each of the identities

$$(32) \qquad (x_u | \zeta) = 0, \qquad (x_v | \zeta) = 0$$

with respect to u and v, we obtain the relations

$$(33) \qquad \begin{aligned} (x_{uu} | \zeta) + (x_u | \zeta_u) &= 0, & (x_{uv} | \zeta) + (x_v | \zeta_u) &= 0, \\ (x_{uv} | \zeta) + (x_u | \zeta_v) &= 0, & (x_{vv} | \zeta) + (x_v | \zeta_v) &= 0. \end{aligned}$$

We note, as an important consequence of these relations, that $(x_u | \zeta_v)$ and $(x_v | \zeta_u)$ are always equal in value.

Differentiating $(\zeta | \zeta) = 1$, we obtain the identities

$$(34) \qquad (\zeta_u | \zeta) = 0, \qquad (\zeta_v | \zeta) = 0.$$

Since these identities say that the vectors ζ_u and ζ_v at P: (u, v) lie in the tangent plane at P, it follows that $\widehat{\zeta_u\,\zeta_v} = k\zeta$. Taking the inner product of each side of this equation with ζ, we find that $k = (\zeta\,\zeta_u\,\zeta_v)$. Hence,

$$(35) \qquad \widehat{\zeta_u\,\zeta_v} = (\zeta\,\zeta_u\,\zeta_v)\zeta,$$

and therefore

$$(36) \qquad (\widehat{\zeta_u\,\zeta_v} | \widehat{\zeta_u\,\zeta_v}) = (\zeta\,\zeta_u\,\zeta_v)^2.$$

Second fundamental form. The first fundamental differential quadratic form of S is, as we have seen, $(dx | dx)$. The second is $-(dx \mid d\zeta)$. We shall denote it by $edu^2 + 2fdudv + gdv^2$:

$$(37) \qquad -(dx \mid d\zeta) \equiv edu^2 + 2fdudv + gdv^2.$$

Since $dx = x_u du + x_v dv$, $d\zeta = \zeta_u du + \zeta_v dv$, we find, with the help of (33) and (30), the following values for e, f, g:

$$(38) \qquad \begin{aligned} e &= -(x_u \mid \zeta_u) = (x_{uu} \mid \zeta) = \frac{1}{D}(x_{uu}\, x_u\, x_v), \\ f &= \begin{Bmatrix} -(x_u \mid \zeta_v) \\ -(x_v \mid \zeta_u) \end{Bmatrix} = (x_{uv} \mid \zeta) = \frac{1}{D}(x_{uv}\, x_u\, x_v), \\ g &= -(x_v \mid \zeta_v) = (x_{vv} \mid \zeta) = \frac{1}{D}(x_{vv}\, x_u\, x_v). \end{aligned}$$

The discriminant $eg - f^2$ of (37) we shall denote by d^2:

$$(39) \qquad d^2 = eg - f^2,$$

even though it may, as we shall see later, be negative. Employing (38), (35), and (31) in turn, we find that

$$(40) \quad d^2 = (x_u\, x_v \mid \zeta_u\, \zeta_v) = (\zeta\, x_u\, x_v)(\zeta\, \zeta_u\, \zeta_v) = D(\zeta\, \zeta_u\, \zeta_v).$$

Order of contact of tangent plane. In determining the order of contact of the tangent plane at P: (u, v) with the surface S, we have to do with the infinitesimal distance from the tangent plane to a point P' on S neighboring to P. The order of this infinitesimal distance depends on the curve C along which P' approaches P. For example, if S is a cylinder, the distance is always zero if P' approaches P along the ruling of the cylinder through P, and otherwise is, in general, not zero.

Accordingly, we shall think of P' as approaching P along a definite curve C through P and shall discuss the order of the distance from P' to the tangent plane at P with respect to the arc PP' of C as principal infinitesimal.

Let C be represented parametrically in terms of its arc s: $u = u(s)$, $v = v(s)$, and denote the arc PP' by Δs. Then P' has the curvilinear coordinates $(u + \Delta u, v + \Delta v)$, where Δu, Δv are the increments of u and v due to Δs. By Taylor's series for a function of two variables, we obtain, as the space coordinates (y_1, y_2, y_3) of P':

$$y = x + x_u \Delta u + x_v \Delta v + \frac{1}{2!}(x_{uu}\Delta u^2 + 2\, x_{uv} \Delta u \Delta v + x_{vv} \Delta v^2) + \cdots,$$

where $x, x_u, x_v, x_{uu}, \cdots$ are evaluated for P.

Inasmuch as $(X - x|\zeta) = 0$ is the equation of the tangent plane at P, $D = (y - x|\zeta)$ is a directed distance from P' to this tangent plane. Since $(x_u|\zeta) = 0$ and $(x_v|\zeta) = 0$,

$$(41)\quad D = \frac{1}{2!}\left[(x_{uu}|\zeta)\Delta u^2 + 2(x_{uv}|\zeta)\Delta u\Delta v + (x_{vv}|\zeta)\Delta v^2\right] + \cdots.$$

Hence

$$\lim_{\Delta s \to 0} \frac{2D}{\Delta s^2} = e\left(\frac{du}{ds}\right)^2 + 2f\frac{du}{ds}\frac{dv}{ds} + g\left(\frac{dv}{ds}\right)^2,$$

and

$$(42)\quad 2D = \left[e\left(\frac{du}{ds}\right)^2 + 2f\frac{du}{ds}\frac{dv}{ds} + g\left(\frac{dv}{ds}\right)^2\right]\Delta s^2 + \epsilon\Delta s^2,$$

where ϵ is an infinitesimal.

Whether or not the expression in the square bracket vanishes depends on the value of the ratio of dv/ds to du/ds, that is, on the value of dv/du. But this is precisely the direction of C at P. Thus, the order of the infinitesimal D depends only on the direction of C at P. In other words, the order of contact of the tangent plane at P depends simply on the direction of approach to P.

THEOREM 1. *The tangent plane at P has in every direction contact of at least the second order if and only if e, f, and g all vanish at P.*

For, it follows from (42) that D is of at least the third order for every direction of departure dv/du if and only if $e = f = g = 0$ at P.

If e, f, and g are not all zero at P, the quadratic equation obtained by setting the expression in the square bracket equal to zero has two, one, or no real roots according as its discriminant, d^2, is negative, zero, or positive.

THEOREM 2. *If e, f, and g are not all zero at P, the tangent plane at P has contact of higher order than the first in two, one, or no real directions, according as $d^2 < 0$, $d^2 = 0$, or $d^2 > 0$ at P.*

All three cases occur. They are illustrated respectively by a point on a ruled quadric surface, say, a hyperboloid of one sheet; by a point on a cylinder; and, by a point on a quadric without real rulings, for example, a sphere.

It is to be noted that we have proved, incidentally, that d^2 may be positive, negative, or zero.

Returning to (42), we remark that, since s is the independent variable throughout, its differential ds is always equal to its increment Δs. Hence, we have

$$(43) \qquad 2D = edu^2 + 2fdudv + gdv^2 + \epsilon\Delta s^2.$$

Consequently, the second fundamental form is, to within terms of higher order, twice the directed distance from P' to the tangent plane at P.

37. Classification of surfaces. Surfaces are classified according to the number of distinct tangent planes. According as a surface has a single tangent plane, a one-parameter family of distinct tangent planes, or a two-parameter family of distinct tangent planes, it is known as a plane, a developable surface, or an ordinary surface.

In seeking criteria for the different types of surfaces, we shall make use of the *spherical representation* of an arbitrary surface S. Imagine all the normal vectors to S translated so that their initial points come to coincide with the origin, O, of coordinates. The locus of their terminal points will, then, lie on the unit sphere, that is, the sphere with center at O and radius unity. This locus is the Gauss, or spherical, representation of S.

Since $\zeta_1(u, v)$, $\zeta_2(u, v)$, $\zeta_3(u, v)$ are the components of the normal vector to S at P: (u, v), they are the coordinates of the terminal point of this vector after it has been translated as described. Hence, the equations

$$(44) \qquad \zeta = \zeta(u, v),$$

where $(\zeta_1, \zeta_2, \zeta_3)$ is now a point, define the spherical representation of S.

According as S is a plane, a developable, or an ordinary surface, the spherical representation is evidently a point, a curve, or a surface. The converse is true, as we shall proceed to show.

Suppose, first, that the spherical representation is a point, and that the coordinate axes have been chosen so that this is the point $(0, 0, 1)$. Then $\zeta_1 = 0$, $\zeta_2 = 0$, or

$$\frac{\partial x_2}{\partial u}\frac{\partial x_3}{\partial v} - \frac{\partial x_2}{\partial v}\frac{\partial x_3}{\partial u} = 0, \quad \frac{\partial x_1}{\partial u}\frac{\partial x_3}{\partial v} - \frac{\partial x_1}{\partial v}\frac{\partial x_3}{\partial u} = 0.$$

The determinant of the coefficients of these two homogeneous linear equations in $\partial x_3/\partial v$, $\partial x_3/\partial u$ is a multiple, not zero, of ζ_3, and $\zeta_3 \neq 0$. Hence, the only solution is $\partial x_3/\partial u = 0$, $\partial x_3/\partial v = 0$. Consequently, $x_3 =$ const., and S is a plane.

Assume, next, that the spherical representation is a curve and that u, v have been so chosen that its parametric equations are of the form $\zeta = \zeta(v)$. Then the normals to S at the points of an arbitrary u-curve are all parallel, and it ought not to be hard to show that the tangent planes at these points are all the same, and hence that S is a developable surface. As a matter of fact, the tangent plane, $(X - x \mid \zeta) = 0$ or $(X \mid \zeta) - (x \mid \zeta) = 0$, is the same all along an arbitrary curve $v =$ const. provided $(x \mid \zeta)$ does not involve u. But, since $\zeta = \zeta(v)$, the partial derivative of $(x \mid \zeta)$ with respect to u is equal to $(x_u \mid \zeta)$ and hence is zero. Thus, the proof is complete.

Finally, if the spherical representation is a surface, S certainly has a two-parameter family of distinct tangent planes and so is an ordinary surface.

THEOREM 1. *A surface is a plane, a developable, or an ordinary surface, according as its spherical representation is a point, a curve, or a surface, and conversely.*

By § 32, the spherical representation $\zeta = \zeta(u, v)$ is a point if and only if $\zeta_u = 0$, $\zeta_v = 0$; a curve if and only if $\zeta_u {}^\frown \zeta_v = 0$, but ζ_u, ζ_v are not both identically zero; and a surface if and only if $\zeta_u {}^\frown \zeta_v \neq 0$. These conditions may be expressed in terms of e, f, g, as indicated in the following theorem.

THEOREM 2. *The surface S is a plane if and only if $e = 0$, $f = 0$, $g = 0$; a developable, if and only if $d^2 = 0$, but e, f, g are not all identically zero; and an ordinary surface if and only if $d^2 \neq 0$.*

If $\zeta_u = 0$ and $\zeta_v = 0$, it follows, from (38), that $e = 0, f = 0$, $g = 0$. Conversely, if $e = f = g = 0$, we have, for example, $(\zeta_u \mid x_u) = 0$, $(\zeta_u \mid x_v) = 0$, whence $\zeta_u = k\zeta$. Taking the inner product of each side of this equation with ζ, we have $k = (\zeta \mid \zeta_u)$. Hence, by (34), $k = 0$, and so $\zeta_u = 0$. Similarly, $\zeta_v = 0$. Thus, the first part of the theorem is proved.

If $\zeta_u {}^\frown \zeta_v = 0$, $d^2 = 0$, by (40). Conversely, if $d^2 = 0$, we have, by (40), that $(\zeta\ \zeta_u\ \zeta_v) = 0$ and hence, by (35), that $\zeta_u {}^\frown \zeta_v = 0$. Thus, the basis for the proof of the second part is provided.

The last part of the theorem now follows by exclusion.

Of the ordinary surfaces, the sphere has the simplest properties, and it is worth our while to characterize it analytically.

THEOREM 3. *The surface S is a sphere if and only if $r(u, v)$ exists so that*

$$E = re, \qquad F = rf, \qquad G = rg. \tag{45}$$

Equations equivalent to (45) are

$$\begin{aligned} (x_u + r\zeta_u | x_u) &= 0, & (x_v + r\zeta_v | x_u) &= 0, \\ (x_u + r\zeta_u | x_v) &= 0, & (x_v + r\zeta_v | x_v) &= 0. \end{aligned} \tag{46}$$

For example, the relation $F = rf$ can be rewritten in the two forms, $(x_u | x_v) + r(\zeta_u | x_v) = 0$ and $(x_v | x_u) + r(\zeta_v | x_u) = 0$, and these reduce to $(x_u + r\zeta_u | x_v) = 0$ and $(x_v + r\zeta_v | x_u) = 0$.

Equations (46) say that the vectors $x_u + r\zeta_u$ and $x_v + r\zeta_v$, which are linear combinations of vectors in the tangent plane, are perpendicular to the tangent plane. Hence, the vectors are null vectors and

$$x_u = -r\zeta_u, \qquad x_v = -r\zeta_v. \tag{47}$$

Differentiating the first of these equations with respect to v, and the second with respect to u, and comparing the results, we find that

$$r_v\zeta_u = r_u\zeta_v. \tag{48}$$

The vectors ζ_u and ζ_v are not parallel, since, otherwise, according to (47), $x_u \frown x_v$ would vanish. Hence it follows from (48) that $r_u = 0$, $r_v = 0$ and r is a constant.

From (47) we conclude that $dx = -r\,d\zeta$. Since r is a constant, this equation is readily integrated, and the result is $x - c = -r\zeta$, where c is a triple of constants. Consequently, $(x - c | x - c) = r^2$, and S is a sphere.

Conversely, if S is a sphere with center at M: (c_1, c_2, c_3) and radius a, and the normal vector ζ at P: (u, v) is taken opposite in sense to the directed radius PM, then $x - c = -a\zeta$. Hence $x_u = -a\zeta_u$, $x_v = -a\zeta_v$ and $E = ae$, $F = af$, $G = ag$.

It is to be noted that equations (45) have a more limited content than the equations $e = kE$, $f = kF$, $g = kG$. For, in the latter equations, k may be identically zero, and S is then, not a sphere, but a plane.

These considerations give rise to the following theorem.

THEOREM 4. *The surface S is a sphere or a plane if and only if the second fundamental form is a multiple of the first.*

EXERCISES

1. Find the first and second fundamental forms of the sphere of § 25, and verify the fact that they are proportional.

2. Prove that

$$x_3 = ax_1 + bx_2 + c(x_1 + kx_2)^2 + d(x_1 + kx_2)^3, \qquad c^2 + d^2 \neq 0,$$

is a developable surface. In computing e, f, g, use the last set of formulas in (38).

38. Classification of points on a surface. In § 36, we learned that the directions at the point P: (u, v) of a surface S in which the tangent plane has contact of at least the second order are the directions at P for which

$$edu^2 + 2\,fdudv + gdv^2 = 0. \tag{49}$$

These directions, if they exist, are known as the *asymptotic directions* at P.

A point at which $e = 0$, $f = 0$, $g = 0$ we shall call a *planar* point. A nonplanar point P is said to be *hyperbolic, parabolic,* or *elliptic* according as $d^2 < 0$, $d^2 = 0$, or $d^2 > 0$ at P.

The results of § 36 we may now restate as follows.

THEOREM 1. *At a planar point every direction is an asymptotic direction. At a nonplanar point there are two, one, or no asymptotic directions according as the point is hyperbolic, parabolic, or elliptic.*

The nature of a planar point we shall discuss later. In investigating the properties of nonplanar points, we first establish the following lemma.

LEMMA. *At an elliptic point, $edu^2 + 2\,fdudv + gdv^2$ has in all directions the same sign; at a parabolic point, it is zero in the single asymptotic direction and has in all other directions the same sign; at a hyperbolic point, it is zero in the two asymptotic directions, has one sign for the directions in the one pair of openings determined by the asymptotic directions, and the opposite sign for the directions in the other pair of openings.*

In giving the proof, we assume that g is not zero at the point, and write

$$edu^2 + 2\,fdudv + gdv^2 = (e + 2\,f\lambda + g\lambda^2)du^2,$$

where $\lambda = dv/du$. The sign of the quadratic form is, then, the same as the sign of $e + 2\,f\lambda + g\lambda^2$. But the equation

$$z = e + 2\,f\lambda + g\lambda^2$$

represents in the Cartesian (λ, z)-plane a parabola whose axis is parallel to the axis of z, and $e + 2\,f\lambda + g\lambda^2$ is always of one sign, always of one sign or zero, or varies in sign, according as the parabola fails to cut the λ-axis, is tangent to it, or cuts it in two distinct points, that is, according as $e + 2\,f\lambda + g\lambda^2 = 0$ has no, one, or two real roots. But the real roots of this equation define the asymptotic directions, and there are no, one, or two asymptotic directions according as the given point is elliptic, parabolic, or hyperbolic. Hence, the Lemma is proved.

The distance D,

$$2\,D = edu^2 + 2\,fdudv + gdv^2 + \epsilon\Delta s^2,$$

from a point P', neighboring to P, to the tangent plane at P is a directed distance, positive for points P' on the one side of the tangent plane, and negative for points P' on the other side. But, when P' is sufficiently close to P, the sign of D is given, except along an asymptotic direction, by the sign of the quadratic form. Hence our Lemma guarantees the following conclusions concerning the surface in a restricted neighborhood of the point P, the point itself being excluded.

THEOREM 2. *At an elliptic point the surface lies wholly on one side of the tangent plane; at a parabolic point, it lies on one side of the tangent plane except perhaps in the single asymptotic direction; at a hyperbolic point, it lies partly on the one side, and partly on the other side, of the tangent plane, and pierces the plane along the two asymptotic directions.*

An ellipsoid, for example, consists exclusively of elliptic points; a hyperboloid of one sheet has only hyperbolic points; and a cone, only parabolic points.

A good example of a surface which possesses points of all three types is the torus, the surface obtained by rotating a circle about a line which lies in its plane but does not cut it. The points on the torus which lie outside the cylinder shown in Fig. 18 are elliptic, those which lie within the cylinder are hyperbolic, and those on the two circles in which the cylinder meets the torus are parabolic.

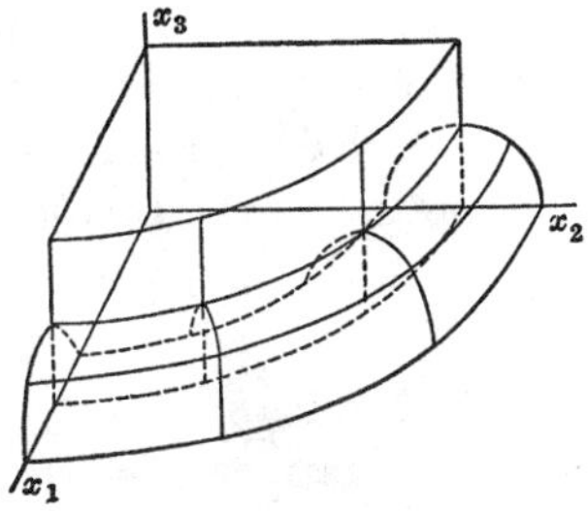

FIG. 18

According to § 37, a surface is a developable if and only if $d^2 = 0$, but e, f, g are not all zero. In other words, the developable surfaces are the only surfaces all of whose points are parabolic. The single asymptotic direction at a point of a developable is that of the ruling through the point.

An elliptic point at which $E = re$, $F = rf$, $G = rg$ (and hence $d^2 = D^2/r^2$) we shall call a *circular* point. By § 37, Theorem 3, the only surfaces all of whose points are circular are the spheres. Hence, at a circular point of a surface, not a sphere, the surface behaves like a sphere; that is, is approximately spherical in shape. The vertex of the paraboloid of revolution obtained by revolving the curve $y = x^2$ about the y-axis is an example of such a point.

Since a planar point is characterized by the vanishing of e, f, g, the only surfaces all of whose points are planar are the planes. An example of a planar point of a surface, other than a plane, is the vertex of the surface obtained by revolving the curve $y = x^4$ about the y-axis. At such a point the surface is approximately flat.

A point at which $e = kE$, $f = kF$, $g = kG$ is either a planar point ($k = 0$) or a circular point ($k \neq 0$). For reasons which will appear presently, a point of this type, whether planar or circular, is known as an *umbilic*.

EXERCISES

1. Show that the upper half of the torus of Fig. 18 is represented by the equations

$$x_1 = u \cos v, \quad x_2 = u \sin v, \quad x_3 = \sqrt{a^2 - (u - b)^2}, \quad b > a > 0,$$

and establish the facts given in the text.

2. Find the umbilics on the surface $x_1x_2x_3 = 1$.

39. Invariant properties of the fundamental forms. *With respect to rigid motions.* Since they have geometrical interpretations which are independent of the position of the surface in space, both fundamental forms are invariant with respect to rigid motions. So also are their coefficients, E, F, G, e, f, g. For, each of these coefficients is the scalar product of two vectors, and the scalar product of two vectors, inasmuch as it can be interpreted as the product of the lengths of the vectors and the cosine of the angle between them, is preserved by every rigid motion.

Behavior with respect to a change of parameters. The parametric equations $x = x(u, v)$, when subjected to the change of parameters

$$u = u(\bar{u}, \bar{v}), \qquad v = v(\bar{u}, \bar{v}),$$

become $x = \bar{x}(\bar{u}, \bar{v})$. Since

$$\bar{x}(\bar{u}, \bar{v}) \equiv x(u, v) \quad \text{where} \quad u = u(\bar{u}, \bar{v}),\, v = v(\bar{u}, \bar{v}),$$

it follows that $d\bar{x} = dx$, and $(d\bar{x}|d\bar{x}) = (dx|dx)$. Hence, the first fundamental form is preserved.

From

(50) $$\bar{x}_{\bar{u}} = x_u u_{\bar{u}} + x_v v_{\bar{u}}, \qquad \bar{x}_{\bar{v}} = x_u u_{\bar{v}} + x_v v_{\bar{v}},$$

we obtain, as the law of transformation of E, F, G,

$$\begin{aligned} \bar{E} &= E u_{\bar{u}}^2 + 2\, F u_{\bar{u}} v_{\bar{u}} + G v_{\bar{u}}^2, \\ \bar{F} &= E u_{\bar{u}} u_{\bar{v}} + F(u_{\bar{u}} v_{\bar{v}} + v_{\bar{u}} u_{\bar{v}}) + G v_{\bar{u}} v_{\bar{v}}, \\ \bar{G} &= E u_{\bar{v}}^2 + 2\, F u_{\bar{v}} v_{\bar{v}} + G v_{\bar{v}}^2. \end{aligned}$$

According to (4),

(51) $$\widehat{\bar{x}_{\bar{u}}\, \bar{x}_{\bar{v}}} = \Delta \widehat{x_u\, x_v}.$$

Hence

(52) $$\bar{D}^2 = \Delta^2 D^2$$

and

(53) $$\bar{D} = \pm\, \Delta D,$$

where the plus or the minus sign is to be taken according as $\Delta > 0$ or $\Delta < 0$.

From (51) and (53), we conclude that

$$\bar\zeta = \pm\,\zeta. \tag{54}$$

Hence, $-(d\bar x\,|\,d\bar\zeta) = \pm\,[-(dx\,|\,d\zeta)]$, that is, the second fundamental form is invariant except perhaps for a change of sign.

Using (50) and

$$\bar\zeta_{\bar u} = \pm\,(\zeta_u u_{\bar u} + \zeta_v v_{\bar u}), \quad \bar\zeta_{\bar v} = \pm\,(\zeta_u u_{\bar v} + \zeta_v v_{\bar v}), \tag{55}$$

we get

$$\begin{aligned} \bar e &= \pm\,(e u_{\bar u}^2 + 2\,f u_{\bar u} v_{\bar u} + g v_{\bar u}^2), \\ \bar f &= \pm\,(e u_{\bar u} u_{\bar v} + f(u_{\bar u} v_{\bar v} + v_{\bar u} u_{\bar v}) + g v_{\bar u} v_{\bar v}), \\ \bar g &= \pm\,(e u_{\bar v}^2 + 2\,f u_{\bar v} v_{\bar v} + g v_{\bar v}^2). \end{aligned}$$

From (55) we obtain the relation

$$\widehat{\bar\zeta_{\bar u}\,\bar\zeta_{\bar v}} = \Delta\,\widehat{\zeta_u\,\zeta_v},$$

which, combined with (51), yields

$$\bar d^2 = \Delta^2 d^2. \tag{56}$$

Finally, we have, from (52) and (56),

$$\bar d^2/\bar D^2 = d^2/D^2. \tag{57}$$

Thus, the expression d^2/D^2, which we know is preserved by rigid motions, is also preserved by every change of parameters. Its geometrical significance will be considered in the next chapter.

The content of equations (52) and (56) we express by saying that D^2 and d^2 are *relative* invariants of weight two with respect to a change of parameters. In distinction, we call d^2/D^2 an *absolute* invariant.

According to (54), the sense of the normal vector is preserved or reversed according as $\Delta > 0$ or $\Delta < 0$. Hence, according as the Jacobian of the change of parameters is positive or negative at a point, the direction of rotation about the point from the positive direction of the first parametric curve to that of the second is preserved or reversed by the change of parameters; see § 36. In the first case, the plus sign is to be taken in the foregoing formulas, and, in the second case, the minus sign.

CHAPTER V

CURVATURE. IMPORTANT SYSTEMS OF CURVES

40. Curvature of a curve on the surface. Let there be given on the surface S: $x = x(u, v)$ a curve C: $u = u(s)$, $v = v(s)$, not a straight line. In order to apply the theory of a space curve to C, we have merely to think of C represented, as a curve in space, by the symbolic equation $x = x(u(s), v(s))$. We have, in particular, at an arbitrary point P: (u, v) of C, the relations $\alpha = dx/ds$ and $\beta/R = d\alpha/ds$. Taking the inner product of each side of the latter equation with ζ, we obtain, as an expression for the curvature, $1/R$, of C at P,

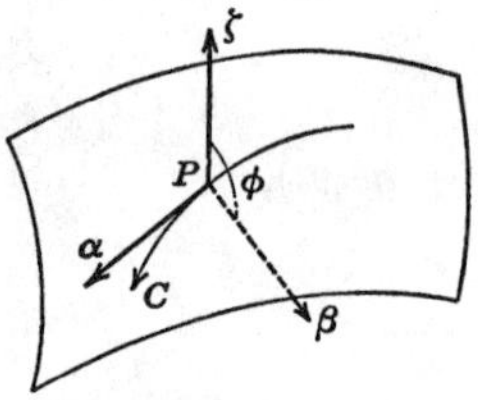

FIG. 19

$$\frac{\cos\phi}{R} = \left(\frac{d\alpha}{ds}\middle|\,\zeta\right),$$

where ϕ is the angle between the principal normal vector, β, to C at P and the normal vector, ζ, to S at P. Inasmuch as $(\alpha\,|\,\zeta) = 0$,

$$\left(\frac{d\alpha}{ds}\middle|\,\zeta\right) + \left(\alpha\,\middle|\,\frac{d\zeta}{ds}\right) = 0.$$

Consequently, since $\alpha = dx/ds$, we have

$$\frac{\cos\phi}{R} = -\left(\frac{dx}{ds}\middle|\,\frac{d\zeta}{ds}\right) = -\frac{(dx\,|\,d\zeta)}{(dx\,|\,dx)},$$

or

$$(1) \qquad \frac{\cos\phi}{R} = \frac{edu^2 + 2\,fdudv + gdv^2}{Edu^2 + 2\,Fdudv + Gdv^2}.$$

If the direction dv/du of the tangent line to C at P is an asymptotic direction, (1) fails to define $1/R$. For, when $edu^2 + 2\,fdudv + gdv^2 = 0$, then it is $\cos\phi$, and not $1/R$, which usually vanishes, as we shall see later. Accordingly, we exclude this case for the present.

When the direction of C at P is not asymptotic, $\cos \phi$ cannot vanish and (1) determines a value, not zero, for $1/R$. Since E, F, G, e, f, g are evaluated at P, and dv/du and $\cos \phi$ depend only on the positions of the tangent and principal normal to C at P, the value of $1/R$ depends only on the position of the osculating plane to C at P. In other words:

THEOREM 1. *If two curves on S passing through a point P have the same osculating plane at P and their common direction at P is not an asymptotic direction, they have the same curvature at P.*

The theorem says, in particular, that the curvature at P of the given curve C is equal to the curvature at P of the plane curve in which the osculating plane of C at P cuts S. Consequently, we may restrict ourselves to plane sections of S.

Consider the sections of S by the planes through the tangent line PT which has the given, nonasymptotic direction dv/du. Among these sections, there is one whose plane is normal to the tangent plane at P. For this normal section,

$$\phi = 0 \qquad \text{or} \qquad \phi = \pi,$$

according as its principal normal at P has the same sense as ζ or the opposite sense. Hence the curvature, $1/R_n$, of the normal section at P is

$$\text{(2)} \qquad \frac{1}{R_n} = \pm \frac{edu^2 + 2\,fdudv + gdv^2}{Edu^2 + 2\,Fdudv + Gdv^2},$$

where the plus sign is to be taken in the first case, and the minus sign, in the second.

For an arbitrary one of the sections under consideration, (1) becomes, in the two cases,

$$\text{(3)} \qquad R = R_n \cos \phi, \qquad R = -\,R_n \cos \phi.$$

As the plane of the section, starting from the normal position, rotates about PT toward the tangential position, the angle ϕ varies continuously. Thus we have, in the two cases,

$$\text{(4)} \qquad 0 \leqq \phi < \frac{\pi}{2}, \qquad \pi \geqq \phi > \frac{\pi}{2},$$

inasmuch as, for the normal section, $\phi = 0$ in the first case, and $\phi = \pi$ in the second.

The content of formulas (3) and (4) is represented in the two cases by Figs. 20a and 20b. Hence, in both cases, the center of curvature of the variable section is the projection, on its principal normal, of the center of curvature, C_n, of the normal section. This result, due to Meusnier, may be stated as follows:

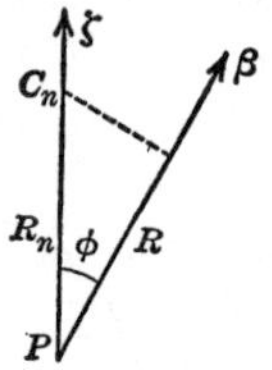

FIG. 20a

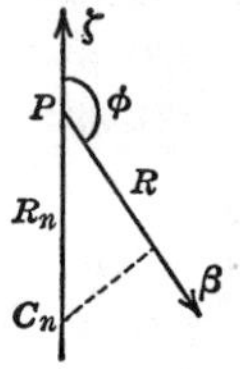

FIG. 20b

THEOREM 2. *If P is a point of a curve C on S at which the direction of C is nonasymptotic, the center of curvature of C at P is the projection, on the principal normal to C at P, of the center of curvature of the normal section which is tangent to C at P.*

It follows that the locus of the center of curvature of the variable section through the tangent line PT is a circle described on the line-segment PC_n as diameter. Moreover, the circle of curvature of the section is the circle in which its plane cuts the sphere whose center is at C_n and whose radius is R_n.

To combine the results of the two foregoing cases in a single set of formulas, we attach to the curvature $1/R_n$ of the normal section through PT a plus sign in the first case, and a minus sign in the second case, and call the signed curvature the *normal curvature of S at P in the direction dv/du.* If we denote this normal curvature by $1/r$, we have in both cases

$$(5) \qquad \frac{1}{r} = \frac{edu^2 + 2\,fdudv + gdv^2}{Edu^2 + 2\,Fdudv + Gdv^2} = -\frac{(dx\,|\,d\zeta)}{(dx\,|\,dx)},$$

for, in the first case, $1/r = 1/R_n$, and, in the second, $1/r = -1/R_n$, so that (2) becomes (5).

The center of curvature, C_n, of the normal section through PT evidently has, in both cases, the coordinates $x + r\zeta$; for, in the first case, $r = R_n$, and, in the second, $r = -R_n$, and Figs. 20a and 20b guarantee, then, the result. The point C_n is called *the center of normal curvature*, and r, *the radius of normal curvature*, in the direction dv/du.

Curves in an asymptotic direction. If the direction of the curve C at P is an asymptotic direction, equation (1) says that $\cos\phi = 0$ or $1/R = 0$, that is, that the osculating plane to C at P is the tangent plane to S at P, or that P is a singular point of C.

The first of these cases is evidently the general one. The very definition of an asymptotic direction implies that the tangent plane to S at P has contact of at least the second order with C. Consequently, the tangent plane is the osculating plane to C at P and $\cos \phi = 0$, except perhaps when P is a singular point of C.

Thus, most of the curves through P in a given asymptotic direction have the same osculating plane at P, whereas their curvatures at P are unrelated. The remaining curves all have the same curvature, zero, at P, while their osculating planes at P are unrelated.

Theorems 1 and 2 do not hold, then, for curves in an asymptotic direction. In particular, they break down completely when S is a plane, inasmuch as every direction at every point of a plane is an asymptotic direction.

The case of a straight line on S, which we have thus far excluded, belongs here. Since the tangent plane to S at a point of the line certainly contains the line, the direction of the line is, at every point on it, an asymptotic direction.

We have seen that, if C is a curve, not a straight line, whose direction at P is asymptotic and whose osculating plane at P is not the tangent plane to S, the curvature of C at P is zero. Hence, a normal section at P in an asymptotic direction always has zero curvature at P. Accordingly, we should, and do, take zero as the value of the normal curvature, $1/r$, in an asymptotic direction. This is in agreement with formula (5) and this formula is, then, universally valid.

41. Normal curvature. In the previous section, we found that the curvature at a point P of an arbitrary curve on the surface S is, in general, simply related to the normal curvature at P in the direction of the curve. To complete the treatment of curvature, it remains to study the law of change of the normal curvature at P in a variable direction.

The normal curvature at P in the direction dv/du is given by

$$\frac{1}{r} = \frac{edu^2 + 2\,fdudv + gdv^2}{Edu^2 + 2\,Fdudv + Gdv^2}, \tag{6}$$

or, if we set $dv/du = \lambda$, by

$$\frac{1}{r} = \frac{e + 2\,f\lambda + g\lambda^2}{E + 2\,F\lambda + G\lambda^2}, \tag{7}$$

where E, F, G, e, f, g are evaluated for P.

We shall consider first the special case in which the normal curvature at P is the same in every direction. It is evident from (7) that this occurs when and only when $e = kE, f = kF, g = kG$ at P. The point P is then an umbilical point or an umbilic (§ 38), and is, in particular, a circular point if $k \neq 0$ and a planar point if $k = 0$. According to § 37, Theorem 4, *a surface all of whose points are umbilics is a sphere or a plane.*

In considering the general case in which e, f, g are not a multiple of E, F, G, we shall assume, in particular, that $Fg - Gf \neq 0$, and plot the curve which represents $1/r$ as a function of λ,—the curve (7). Since the denominator in (6), because of its geometrical significance, is always positive, the denominator in (7) is never zero, and the curve in question is continuous for all values of λ. Moreover, the curve is cut by a line parallel to the λ-axis in at most two (real) points; for, when $1/r$ is given, a quadratic equation is obtained for the determination of λ.

When λ becomes infinite, either positively or negatively, $1/r$ approaches g/G as a limit. Hence, the curve has a horizontal asymptote which it approaches in both directions. Furthermore, it cuts this asymptote at a finite point; for, when we set g/G for $1/r$ in (7), we get $2(Fg - Gf)\lambda = Ge - Eg$, and $Fg - Gf \neq 0$, by hypothesis.

It follows, now, that the curve has the character of either the unbroken or the dotted graph in Fig. 21. Consequently, the normal curvature has two extrema, an absolute maximum and an absolute minimum.

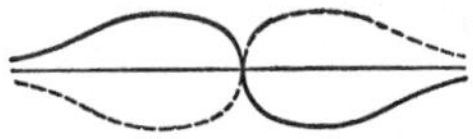

FIG. 21

The two directions in which the normal curvature has its extrema are known as the *principal directions* at the point P. Since they are the directions for which the derivative of $1/r$ with respect to λ vanishes, it is readily shown that the values of λ which define them are the roots of the quadratic equation

$$(Ef - Fe) + (Eg - Ge)\lambda + (Fg - Gf)\lambda^2 = 0.$$

Thus, *the principal directions at the point P are defined by the equation*

$$(8) \quad (Ef - Fe)du^2 + (Eg - Ge)du\,dv + (Fg - Gf)dv^2 = 0,$$

where E, F, G, e, f, g are evaluated for P.

THEOREM 1. *The principal directions at a point are mutually perpendicular.*

This important fact is easily verified by applying Theorem 2 of § 35 to equation (8).

The two extreme values of the normal curvature are called the *principal normal curvatures* at the point P. They are the values of $1/r$ for which the two corresponding values of λ, as determined from (7), are equal. But the quadratic equation in λ obtained from (7), namely

$$\left(\frac{E}{r} - e\right) + 2\left(\frac{F}{r} - f\right)\lambda + \left(\frac{G}{r} - g\right)\lambda^2 = 0,$$

has equal roots if and only if its discriminant vanishes. Thus, we obtain, as the equation in $1/r$ whose roots are the principal normal curvatures,

$$(9) \quad (EG - F^2)\frac{1}{r^2} - (Eg - 2\,Ff + Ge)\frac{1}{r} + (eg - f^2) = 0.$$

We shall henceforth denote the principal normal curvatures by $1/r_1$, $1/r_2$. We shall, however, be interested, not so much in these curvatures themselves, as in their product and sum, which we shall designate respectively by K and K':

$$(10) \qquad K = \frac{1}{r_1}\frac{1}{r_2}, \qquad K' = \frac{1}{r_1} + \frac{1}{r_2}.$$

The product K is called the *total curvature*, or the *Gaussian curvature*, of the surface at P, and the sum K' is known as the *mean curvature* of the surface at P.

Since $1/r_1$, $1/r_2$ are the roots of (9), we have

$$(11) \qquad K = \frac{d^2}{D^2}, \qquad K' = \frac{Eg - 2\,Ff + Ge}{D^2}.$$

In § 39, we noted that D^2 and d^2 are relative invariants of weight two with respect to a change of parameters. It follows now, since K' is surely an absolute invariant, that $Eg - 2\,Ff + Ge$ is also a relative invariant of weight two. It is known as the simultaneous (relative) invariant of the two fundamental forms.

The foregoing results are independent of the initial assumption that $Fg - Gf \neq 0$. Since P is not an umbilic, at least one of the quantities $Fg - Gf$, $Ef - Fe$, $Eg - Ge$ is not zero. If $Fg - Gf$ were zero, but $Ef - Fe$ not zero, we would set $du/dv = \mu$ in (6) and proceed precisely as before. If $Fg - Gf$ and $Ef - Fe$ were both zero, it would follow that $F = 0, f = 0$. The graph of $1/r$ as a function of λ would then be somewhat changed in character, but the general procedure would remain the same.

The principal directions at P have been defined only when P is not an umbilic. If P is an umbilic, $1/r$ is the same in all directions. Moreover, the coefficients in (8) are all zero. Thus, it is natural to think of every direction at an umbilic as a principal direction. Every direction at a point of a sphere or a plane is, then, a principal direction.

As the common value of the principal normal curvatures at an umbilic, P, we take the constant value of $1/r$ at P. As a matter of fact, this is the result which (9) would give in this case. For, if $1/a$ is the constant value of $1/r$, then $e = E/a$, $f = F/a$, $g = G/a$, and both roots of (9) reduce to $1/a$.

It follows that, at the umbilic P, $K = 1/a^2$. In the case of the sphere, $1/a$ is evidently equal numerically to the curvature of a great circle and so is the same for each point P. Thus, the total curvature of a sphere is constant and positive. In the case of a plane, $1/a = 0$: the total curvature of a plane is zero.

Applications. In § 40, we learned that the asymptotic directions at P are the directions at P in which the normal curvature is zero. In other words, the normal section at P in a given direction behaves at P like a straight line if and only if the direction is an asymptotic direction.

The fact that $K = d^2/D^2$ has the same sign as d^2 may be interpreted as follows:

THEOREM 2. *A nonplanar point P is an elliptic, parabolic, or hyperbolic point according as $K > 0$, $K = 0$, or $K < 0$ at P.*

We are now in a position to prove anew Theorem 2 of § 38. If P is an elliptic point, $K > 0$ and $1/r_1$, $1/r_2$ are of the same sign. Then $1/r$ is always of this sign, since $1/r$ varies between $1/r_1$ and $1/r_2$. Hence, the centers of curvature of the normal sections at P all lie on one half of the surface normal, and therefore the surface in the neighborhood of P lies on one side of the tangent plane.

If P is a hyperbolic point, $K < 0$ and $1/r_1$, $1/r_2$ are opposite in sign. In this case, $1/r$ is positive for some directions, negative for others, and zero in the two asymptotic directions. Consequently, the surface lies in part on the one side, and in part on the other side, of the tangent plane, and cuts through it along the asymptotic directions.

Finally, if P is a parabolic point, $K = 0$ and either $1/r_1 = 0$ or $1/r_2 = 0$. In other words, the single asymptotic direction coincides with a principal direction. Except perhaps along this direction, the surface lies on one side of the tangent plane.

If every point of a surface is an elliptic point, K is always positive and the surface is known as a *surface of positive curvature.* An ellipsoid is a surface of positive curvature, and a sphere is a surface of constant positive curvature.

A hyperbolic paraboloid is an example of a *surface of negative curvature,* that is, a surface all of whose points are hyperbolic.

If $K = 0$, that is, if all the points of a surface are parabolic or planar, then $d^2 = 0$, and conversely. Hence, by § 37, Theorem 2, we pass to the following conclusion.

THEOREM 3. *The only surfaces for which the total curvature is identically zero are the developables and the planes.*

EXERCISES

1. Prove that, if at every point of a surface a principal direction coincides with an asymptotic direction, the surface is a developable or a plane.

2. Show that an ordinary surface is a sphere if and only if $K'^2 = 4K$.

42. Euler's equation. *Lines of curvature.* A curve on a surface whose direction at each and every point is a principal direction is known as a line of curvature. If the surface is not a sphere or a plane, there are two principal directions at each point, other than an umbilic, and they are mutually perpendicular. Hence, in this case, there are two families of lines of curvature, and they form an orthogonal system. According to (8),

$$(12)\qquad (Ef - Fe)du^2 + (Eg - Ge)dudv + (Fg - Gf)dv^2 = 0$$

is the differential equation of this system.

Every curve on a sphere or a plane is a line of curvature, since, in these cases, every direction at a point is a principal direction.

However, by a system of lines of curvature on a sphere or a plane, we shall always understand an orthogonal system.

THEOREM 1. *A necessary and sufficient condition that the system of parametric curves consist of lines of curvature is that $F = 0$ and $f = 0$.*

The theorem is true for a sphere or a plane, since, in these cases, $e = kE$, $f = kF$, $g = kG$, so that, whenever $F = 0$, then $f = 0$ also.

For a surface, other than a sphere or a plane, equation (12) reduces, when $F = 0$ and $f = 0$, to $dudv = 0$, and hence the lines of curvature are the parametric curves. Conversely, if (12) reduces to $dudv = 0$, then $Ef - eF = 0$ and $Gf - gF = 0$. But, since $Eg - Ge \neq 0$, the only solution of these two equations in f and F is $f = 0$, $F = 0$. Thus, the theorem is established.

Euler's equation. We assume that the parametric curves are lines of curvature: $F = 0$, $f = 0$. Then

$$\frac{1}{r} = \frac{edu^2 + gdv^2}{Edu^2 + Gdv^2}. \tag{13}$$

Since the directions of the parametric curves at a point are now principal directions at the point, the principal normal curvatures are obtained by setting in turn $dv = 0$ and $du = 0$ in (13). Consequently, if $1/r_1$ is the principal normal curvature in the direction of the u-curve, and $1/r_2$ that in the direction of the v-curve,

$$\frac{1}{r_1} = \frac{e}{E}, \quad \frac{1}{r_2} = \frac{g}{G}.$$

We now write (13) in the form

$$\frac{1}{r} = \frac{1}{r_1}\frac{Edu^2}{Edu^2 + Gdv^2} + \frac{1}{r_2}\frac{Gdv^2}{Edu^2 + Gdv^2},$$

and note that

$$\frac{Edu^2}{Edu^2 + Gdv^2} = \cos^2\theta, \qquad \frac{Gdv^2}{Edu^2 + Gdv^2} = \sin^2\theta, \tag{14}$$

where θ is the angle from the positive direction of the u-curve to the direction dv/du. Hence

$$\frac{1}{r} = \frac{\cos^2\theta}{r_1} + \frac{\sin^2\theta}{r_2}. \tag{15}$$

This equation, which is due to Euler, expresses the normal curvature in a given direction at a point P in terms of the principal normal curvatures at P and the angle through which one of the principal directions at P must be rotated in order that it coincide with the given direction. Inasmuch as all of these quantities pertain only to the surface and are independent of the particular systems of axes and parametric curves to which the surface is referred, the equation is an intrinsic or natural equation. It portrays the behavior of the normal curvature at P in the simplest and most striking manner possible.

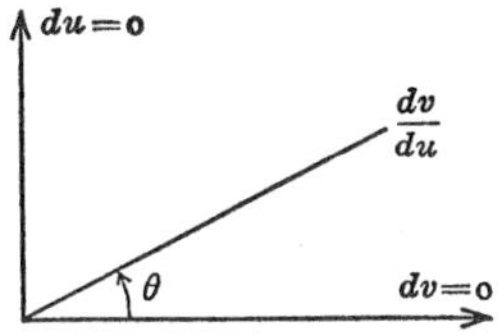

FIG. 22

Euler's equation tells us that $1/r$ has the same value for two angles θ which are negatives of one another. Hence, *the normal curvatures at P in two directions which are symmetrically situated with respect to a principal direction are equal.*

If P is not an umbilic, we know that there are two directions in which the normal curvature has a prescribed value between $1/r_1$ and $1/r_2$. It is now clear that these two directions are equally inclined to each principal direction. In particular:

THEOREM 2. *The asymptotic directions at a hyperbolic point are equally inclined to each principal direction at the point.*

The previous result, to the effect that the single asymptotic direction at a parabolic point coincides with a principal direction, follows also from (15).

The asymptotic directions at a hyperbolic point are perpendicular if and only if $1/r = 0$ when $\theta = \pm \pi/4$, that is, when and only when $1/r_1 + 1/r_2 = 0$. In other words:

THEOREM 3. *A necessary and sufficient condition that the asymptotic directions at a hyperbolic point be mutually perpendicular is that the mean curvature, K', vanish at the point.*

A surface, other than a plane, for which K' vanishes identically is known as a *minimal surface*. It is a surface of negative curvature, at every point of which the asymptotic directions cut orthogonally.

THEOREM 4. *The sum of the normal curvatures at a point P in two perpendicular directions is the same for each two perpendicular directions, and is equal to the mean curvature at P.*

The theorem follows readily from Euler's equation.

Example. A *right helicoid* is a ruled surface which is generated by a variable line which always intersects a fixed line at right angles and rotates about, and slides along, the fixed line so that its points trace circular helices. It has the appearance of a double winding staircase without steps, —the circular ramp of a motor mart.

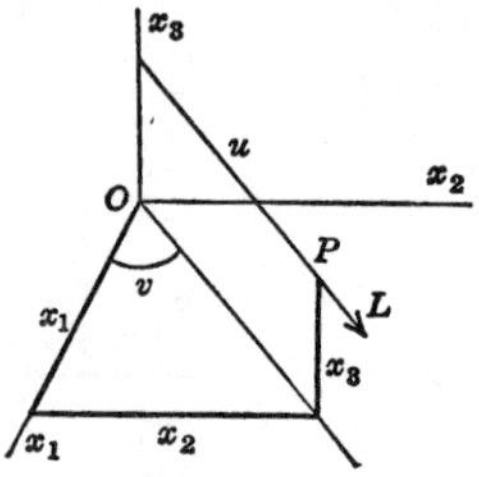

FIG. 23

If we take the axis of x_3 as the fixed line and the axis of x_1 as one of the positions of the variable line, parametric equations of the right helicoid are

$$(16) \qquad x_1 = u\cos v, \qquad x_2 = u\sin v, \qquad x_3 = av, \qquad a \neq 0,$$

where u is the directed distance on an arbitrary ruling, L, measured from the point in which L meets the x_3-axis, and v is the angle which L makes with the positive axis of x_1 (Fig. 23). The circular helices on the helicoid are the curves u = const. and the rulings are the curves v = const.

The components of x_u and x_v are respectively $\cos v$, $\sin v$, 0, and $-u\sin v$, $u\cos v$, a; therefore,

$$E = 1, \quad F = 0, \quad G = u^2 + a^2, \quad D^2 = u^2 + a^2.$$

Since $F = 0$, the parametric curves, the rulings and the circular helices, form an orthogonal system. As a matter of fact, we know, from § 15, that the rulings are the common principal normals of the helices.

For x_{uu}, x_{uv}, x_{vv} we find the components 0, 0, 0; $-\sin v$, $\cos v$, 0; $-u\cos v$, $-u\sin v$, 0. Hence, the values of e, f, g, computed by means of the last set of formulas in equations (38) of Chapter IV, are

$$e = 0, \quad f = -\,a/D, \quad g = 0, \quad d^2 = -\,a^2/D^2.$$

Since $e = 0$ and $g = 0$, the asymptotic directions at P: (u, v), as given by equation (49) of Chapter IV, are the directions $dv = 0$ and $du = 0$ at P, that is, the directions of the ruling and the circular helix passing through P. At every point P these directions are mutually perpendicular, and therefore the right helicoid is a minimal surface.

Equation (12) becomes, in this case, $du^2 - (u^2 + a^2)dv^2 = 0$. Hence, the differential equations of the two families of lines of curvature on the helicoid are

$$du - \sqrt{u^2 + a^2}\, dv = 0, \qquad du + \sqrt{u^2 + a^2}\, dv = 0.$$

Integration of these equations gives the finite equations of the lines of curvature, namely,

$$\sinh^{-1}(u/a) - v = \text{const.}, \qquad \sinh^{-1}(u/a) + v = \text{const.}$$

By (11), we find that $K = -a^2/(u^2 + a^2)^2$. Hence, the total curvature of the helicoid is the same at all the points of a circular helix, $u = \text{const.}$, and approaches zero when the radius of the circular cylinder containing the helix becomes infinite.

EXERCISES

1. The general ruled surface which is generated by a variable line which always intersects a fixed line at right angles is known as a *right conoid*. It may be represented parametrically by the equations

$$x_1 = u \cos v, \quad x_2 = u \sin v, \quad x_3 = \phi(v).$$

Compute E, F, G, e, f, g, K, K' for the conoid. Show that the only conoid which is a minimal surface is the right helicoid. Prove also that, if the total curvature of the conoid is the same at all the points of an arbitrary orthogonal trajectory of the rulings, the conoid is the right helicoid.

2. Determine the lines of curvature on the torus of § 38, Ex. 1, and identify them geometrically. Find $1/r_1$, $1/r_2$ and show that they are equal numerically to the radius of the generating circle and the distance from the chosen point to the axis, measured along the normal. Hence prove that, if the torus contains "horizontal" circles of radius $b/2$, the mean curvature is zero in the points of these circles and in no other points.

3. What are the lines of curvature on the tangent surface of a space curve? Find the principal normal curvatures and the curvatures of the lines of curvature.

4. Show that the sum of the normal curvatures at a point in two perpendicular directions is the same for each two perpendicular directions.

43. Dupin's indicatrix of the normal curvature. In each direction dv/du at a nonplanar point P, except the asymptotic directions, the normal curvature, $1/r$, is not zero, and the radius, r, of normal curvature exists. We may, then, mark on the tangent line at P in the direction dv/du the two points which are at a distance from P equal to $\sqrt{|r|}$. The locus of these points is a curve which pictures the manner in which $\sqrt{|r|}$ varies with the direction. This curve is known as the Dupin indicatrix of the normal curvature.

The Dupin indicatrix lies in the tangent plane at P. We shall write its equation in terms of Cartesian coordinates (ξ_1, ξ_2) in this plane. As the axes of ξ_1, ξ_2 we take, respectively, the directed tangents to the u-curve and v-curve through P, assuming, as in § 42, that these curves are lines of curvature. Then, the point of the Dupin indicatrix which lies on the radius vector making the angle θ with the ξ_1-axis has the coordinates $\xi_1 = \sqrt{|r|}\cos\theta$, $\xi_2 = \sqrt{|r|}\sin\theta$. Hence, it follows, from (15), that the equation of the Dupin indicatrix is

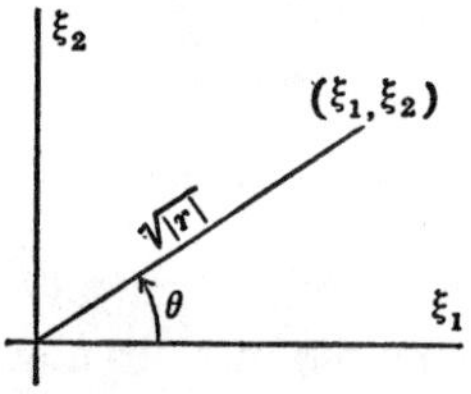

FIG. 24

$$\frac{\xi_1^2}{r_1|r|} + \frac{\xi_2^2}{r_2|r|} = \frac{1}{r}. \tag{17}$$

If P is an elliptic point, r_1, r_2, and r are all of the same sign and hence may be replaced in (17) by $|r_1|$, $|r_2|$, and $|r|$. Then (17) becomes

$$\frac{\xi_1^2}{|r_1|} + \frac{\xi_2^2}{|r_2|} = 1.$$

Thus, the Dupin indicatrix at an elliptic point is an ellipse whose axes coincide with the principal directions and whose semi-axes are $\sqrt{|r_1|}$, $\sqrt{|r_2|}$. If the point is, in particular, circular, the ellipse is a circle.

If P is hyperbolic, r_1 and r_2 are opposite in sign and r may have the sign of either. If r has the same sign as r_1, then r, r_1, and $-r_2$ have the same sign, and (17) becomes

$$\frac{\xi_1^2}{|r_1|} - \frac{\xi_2^2}{|r_2|} = 1.$$

If r has the sign of r_2, (17) reduces to

$$-\frac{\xi_1^2}{|r_1|} + \frac{\xi_2^2}{|r_2|} = 1.$$

Thus, the Dupin indicatrix at a hyperbolic point consists of two conjugate hyperbolas. The directions of the common asymptotes of these hyperbolas, since they are the directions for which $|r|$ becomes infinite, are the asymptotic directions at the point. Hence, the one hyperbola is the indicatrix of the normal curvature for the directions in which the surface lies on one side of the tangent plane, and the other hyperbola represents the normal curvature for the directions in which the surface is on the other side of the tangent plane.

If the point P is parabolic, either $1/r_1$ or $1/r_2$ is zero. Assume, say, that $1/r_1 = 0$. Then r is of the same sign as r_2 and (17) becomes

$$\xi_2^2 = |r_2|.$$

Thus, the Dupin indicatrix at a parabolic point is a degenerate parabola consisting of two distinct straight lines which are parallel to the asymptotic direction at the point.

Form of the surface in the neighborhood of a point. The Dupin indicatrix amounts to something more than an artificial representation of the variation of the normal curvature at a point. It actually represents the form of the surface in the neighborhood of the point. The nature of this representation is described in the following theorem.

THEOREM 1. *If the surface in the neighborhood of a nonplanar point P is cut by a plane which is parallel to the tangent plane at P and near to it, the curve of intersection is approximately a conic which is similar and similarly placed to the Dupin indicatrix at P, or, if P is a hyperbolic point, to one of the hyperbolas constituting the Dupin indicatrix.*

Since the parametric curves are lines of curvature, formula (41), of Chapter IV, for the directed distance D from the point P': $(u + \Delta u, v + \Delta v)$ of the surface to the tangent plane, T, at P: (u, v) becomes

$$2D = e\Delta u^2 + g\Delta v^2 + \cdots. \tag{18}$$

If D is thought of as the fixed directed distance from T to a neighboring plane, M, parallel to T, this equation represents the curve in which the surface in the neighborhood of P is cut by M.

To find the equation of the curve, as a curve in M, we introduce in M Cartesian coordinates (η_1, η_2), taking as the axes of η_1, η_2 the projections on M of the axes of ξ_1, ξ_2 in T. Then, the coordinates η_1 and η_2 of P' are the directed distances from P' to the planes through P which are normal, respectively, to the unit vectors $x_u/\sqrt{E}$ and $x_v/\sqrt{G}$. Consequently

$$\eta_1 = \frac{(y - x \,|\, x_u)}{\sqrt{E}}, \qquad \eta_2 = \frac{(y - x \,|\, x_v)}{\sqrt{G}},$$

where (y_1, y_2, y_3) are the space coordinates of P'. Since

$$y = x + x_u \Delta u + x_v \Delta v + \cdots,$$

and $F = 0$, we find that

$$\eta_1 = \sqrt{E}\,\Delta u + \cdots, \qquad \eta_2 = \sqrt{G}\,\Delta v + \cdots,$$

where the omitted terms are of at least the second degree in Δu, Δv. It follows, then, by a theorem in analysis, that

$$\Delta u = \eta_1/\sqrt{E} + \cdots, \qquad \Delta v = \eta_2/\sqrt{G} + \cdots,$$

where the omitted terms are at least of the second degree in η_1, η_2. Substituting these values for Δu and Δv in (18) and recalling that $e/E = 1/r_1$ and $g/G = 1/r_2$, we obtain, as the desired equation of the curve,

$$(19) \qquad 2\,D = \frac{\eta_1^2}{r_1} + \frac{\eta_2^2}{r_2} + \cdots,$$

where the omitted terms are at least of the third degree in η_1, η_2. Consequently, the curve approximates to the conic $\eta_1^2/r_1 + \eta_2^2/r_2 = 2\,D$, and the theorem is proved.

It follows that the planes which are normal to the surface at P and meet T in the tangents to the lines of curvature at P are, approximately, planes of symmetry of the surface in the immediate neighborhood of P.

44. Lines of curvature. The lines of curvature on a surface, which were defined as the curves whose directions are always principal directions, have a second, equally important, characteristic property, which may be stated as follows.

THEOREM 1. *A curve C on the surface S is a line of curvature if and only if the normals to S at the points of C generate a developable surface or a plane.*

For example, the meridians on the torus, that is, the various positions of the circle whose rotation generates the torus (Fig. 18), are curves with this property. So, also, are the parallels: the circles traced by the points of the generating circle. Thus the lines of curvature on the torus, and, in fact, on any surface of revolution, are the meridians and parallels.

In establishing the theorem, we shall show, first, that a necessary and sufficient condition that the normals to S at the points of $C: u = u(s)$, $v = v(s)$ generate a developable or a plane is that there exists a function, $1/\rho$, of s such that the equation

$$\frac{1}{\rho}\frac{dx}{ds} + \frac{d\zeta}{ds} = 0 \tag{20}$$

is an identity in s.

If the normals to S at the points of C envelope a curve, there exists a function, $\rho(s)$, not zero, so that

$$X = x + \rho\zeta, \tag{21}$$

where $x = x(u(s), v(s))$ and $\zeta = \zeta(u(s), v(s))$, represents this curve, that is, so that $dX/ds = k\zeta$, where $k \neq 0$. Since

$$\frac{dX}{ds} = \frac{dx}{ds} + \rho\frac{d\zeta}{ds} + \frac{d\rho}{ds}\zeta, \tag{22}$$

the relation $dX/ds = k\zeta$ becomes

$$\frac{dx}{ds} + \rho\frac{d\zeta}{ds} + \left(\frac{d\rho}{ds} - k\right)\zeta = 0.$$

Taking the inner product with ζ, we find that $d\rho/ds = k$, and hence that relation (20) holds.

If the normals to S at the points of C go through a point, then $\rho(s)$, not zero, exists so that (21) represents this point, that is, so that $dX/ds = k\zeta$, where $k = 0$. Hence, we arrive at the relation (20), as before. We also conclude that, in this case, ρ is constant.

Finally, if the normals to S at the points of C are all parallel, then $d\zeta/ds = 0$. Evidently, (20) holds here also, for we have merely to take $1/\rho = 0$.

Conversely, let there be given a curve C along which an identity of the form (20) is valid. If $1/\rho = 0$, then $d\zeta/ds = 0$ and the normals to S along C are all parallel. If $1/\rho$ is constant, not zero, it follows from (22) that (21) represents a point, and hence that the normals to S along C go through a point. Finally, if $1/\rho$ is not constant, (21) represents a curve whose tangents are the normals to S along C. Hence, our proposition is proved.

It remains to show that a curve C for which a relation of the form (20) holds is a line of curvature, and conversely. In discussing this question, we shall write (20) more simply as

$$\frac{dx}{\rho} + d\zeta = 0. \tag{23}$$

Expanding this relation so that it reads

$$\frac{1}{\rho}(x_u du + x_v dv) + (\zeta_u du + \zeta_v dv) = 0,$$

and taking the inner product with x_u and x_v in turn, we obtain the equations

$$\begin{aligned} \frac{1}{\rho}(Edu + Fdv) - (edu + fdv) &= 0, \\ \frac{1}{\rho}(Fdu + Gdv) - (fdu + gdv) &= 0. \end{aligned} \tag{24}$$

Eliminating $1/\rho$, we get the equation

$$\begin{vmatrix} Edu + Fdv & edu + fdv \\ Fdu + Gdv & fdu + gdv \end{vmatrix} = 0, \tag{25}$$

which, when expanded, becomes

$$(Ef - Fe)du^2 + (Eg - Ge)dudv + (Fg - Gf)dv^2 = 0.$$

Hence, the curve C is a line of curvature.

Suppose, conversely, that the given curve C is a line of curvature, that is, a curve for which (25) is valid. Then equations (24) in $1/\rho$ are compatible and determine $1/\rho$ uniquely. But equations (24) can be written in the forms

$$\left(\frac{dx}{\rho} + d\zeta \,|\, x_u\right) = 0, \qquad \left(\frac{dx}{\rho} + d\zeta \,|\, x_v\right) = 0,$$

and hence $dx/\rho + d\zeta = c\zeta$. Taking the inner product of both sides with ζ, we find that $c = 0$. Hence, $dx/\rho + d\zeta = 0$ along C.

The proof of Theorem 1 is now complete. Moreover, we have obtained a new analytic criterion for a line of curvature:

THEOREM 2. *A curve is a line of curvature if and only if a function* $1/\rho$ *exists so that along the curve*

$$\frac{dx}{\rho} + d\zeta = 0. \tag{23}$$

It is to be remembered that equation (23) is a simplified form of (20), that the left-hand side is actually a function of the parameter in terms of which the curve is represented and is to vanish identically for all values of this parameter. The equation is due to Olinde Rodrigues.

Taking the inner product of each side of (23) with dx/ds, or, more simply, with dx, we get $1/\rho = -(dx|d\zeta)/(dx|dx)$. Hence, $1/\rho$ *is precisely the principal normal curvature,* $1/r$, *in the direction of the line of curvature.*

This result, combined with the detailed facts concerning $1/\rho$ obtained in the first half of the proof of Theorem 1, leads us to the following conclusions.

THEOREM 3. *According as the principal normal curvature in the direction of a line of curvature is variable, constant but not zero, or zero, the normals to the surface along the line of curvature envelope a curve, go through a point, or are parallel.*

In general, the normals to a surface along a line of curvature actually envelope a curve. According to § 31, this curve is an evolute of the line of curvature.

The theorems of Joachimsthal. Suppose that two surfaces S_1 and S_2 intersect in a curve C which is a line of curvature on both surfaces. Since the normals to S_1 along C and the normals to S_2 along C constitute two one-parameter families of normals to C, both of which form developables or planes, the angle between the two normals at a point of C is the same at all the points of C, by § 31, Theorem 2. But this angle is one of the angles at which the surfaces S_1 and S_2 intersect. Hence, we draw the following conclusion.

THEOREM 4. *If a curve of intersection of two surfaces is a line of curvature on both surfaces, the surfaces intersect at the same angle all along the curve.*

Similarly, we may establish a converse:

THEOREM 5. *If two surfaces intersect at the same angle at all the points of a curve and the curve is a line of curvature on one of the surfaces, it is a line of curvature on the other.*

Recalling that every curve on a sphere or a plane is a line of curvature, we deduce from Theorems 4 and 5 the following interesting result.

THEOREM 6. *A plane or a sphere intersects a surface at the same angle at all the points of a curve if and only if the curve is a line of curvature on the surface.*

Closely related to these theorems is a theorem of Dupin on triply orthogonal systems of surfaces. A system of surfaces is said to be triply orthogonal if it consists of three one-parameter families of surfaces such that (*a*) through each point there passes, in general, just one surface of each family, and (*b*) two surfaces of different families intersect always at right angles. Dupin's theorem, which we shall state without proof, reads as follows:

THEOREM 7. *The curve in which two surfaces of different families of a triply orthogonal system intersect is a line of curvature on both surfaces.*

It is a fact, with which the reader may be familiar, that if a central quadric, Q, is given, there exists a triply orthogonal system of quadric surfaces of which Q is a member. It follows from Dupin's theorem that the two families of lines of curvature on Q are the curves in which Q is met by the quadrics of the two families of the system to which Q does not belong. Hence, a line of curvature on a central quadric is, in general, a twisted quartic, that is, a twisted curve which is met by a plane, in general, in four points, real or imaginary.

EXERCISES

1. Using the results of this section, prove anew the facts concerning the principal normal curvatures of the torus given in § 42, Ex. 2.

2. Show that a curve on a surface is a line of curvature along which the normals go through a point when and only when the curve is spherical and a sphere on which it lies is tangent to the surface all along it. (The meridians and all but two of the parallels on the torus are lines of curvature of this type.)

3. Prove that a curve on a surface is a line of curvature along which the normals are all parallel if and only if the curve is a plane curve and a plane in which it lies is tangent to the surface all along it. (The rulings on a developable surface and two of the parallels on the torus are lines of curvature of this type.)

4. Show that when the osculating plane of a line of curvature makes always the same angle with the tangent plane to the surface, the line of curvature is a plane curve.

5. Show that an orthogonal trajectory of the rulings of a cone is a plane curve if and only if the cone is a cone of revolution.

6. The symbolic equation $x = x(u, v, w)$, where the Jacobian of x_1, x_2, x_3 with respect to u, v, w does not vanish identically, defines three families of surfaces, namely, the surfaces $u =$ const., on each of which v, w serve as curvilinear coordinates, the surfaces $v =$ const., and the surfaces $w =$ const. Show that, if the surfaces form a triply orthogonal system, $(x_u \mid x_v) = 0$, $(x_v \mid x_w) = 0$, $(x_w \mid x_u) = 0$ and therefore $(x_u \mid x_{vw}) = 0$, $(x_v \mid x_{wu}) = 0$, $(x_w \mid x_{uv}) = 0$. Hence prove Dupin's theorem.

45. Conjugate systems of curves. *Conjugate diameters of a central conic.* We recall that a diameter of a central conic, that is, a line through the center, bisects a certain set of parallel chords, and that, if one of two diameters bisects the chords parallel to the other, the second bisects the chords parallel to the first. Two diameters related in this way are said to be conjugate to one another. Two diameters of the conic $x^2/A + y^2/B = \pm 1$ are conjugate if and only if they are the axes of the conic or their slopes, λ and λ', satisfy the equation $\lambda\lambda' = -B/A$.

A parabola has no conjugate diameters. It is true that the mid-points of a set of parallel chords of a parabola lie on a line, but the line is parallel to the axis and the parabola has no chords parallel to the axis.

Conjugate directions. The directions of a pair of conjugate diameters of the Dupin indicatrix at an elliptic or hyperbolic point P of a surface S are known as conjugate directions at P. Since there exists a unique direction which is conjugate to a prescribed direction, there are infinitely many pairs of conjugate directions at P. Evidently, two conjugate directions which are mutually perpendicular are necessarily principal directions; in particular, each two conjugate directions at a circular point are mutually perpendicular.

Conjugate directions are not defined at a parabolic or planar point. Hence, developables and planes are excluded completely.

In deducing an analytic criterion for conjugate directions at an elliptic or hyperbolic point P, we assume, first, that the parametric curves on S are lines of curvature and that the tangent plane at P has been referred to the Cartesian coordinates (ξ_1, ξ_2) employed in § 43. The equation of the Dupin indicatrix at P is, then, $\xi_1^2/r_1 + \xi_2^2/r_2 = r/|r|$. Hence, the two directions dv/du and $\delta v/\delta u$ at P are conjugate, according to the above condition, when and only when

$$\tan\theta \tan\theta' = -\frac{r_2}{r_1}, \tag{26}$$

where θ and θ' are respectively their slope angles referred to the positive direction of the u-curve at P.

From (14), we have

$$\tan\theta = \pm\sqrt{\frac{G}{E}}\frac{dv}{du}, \qquad \tan\theta' = \pm\sqrt{\frac{G}{E}}\frac{\delta v}{\delta u},$$

where the same choice of sign is to be made in both formulas. Hence, since $1/r_1 = e/E$ and $1/r_2 = g/G$, (26) becomes

$$edu\delta u + gdv\delta v = 0. \tag{27}$$

This, then, is the form of the criterion for conjugate directions when the lines of curvature are parametric.

Consider, now, the tangent planes to S at the points of a curve C: $u = u(s)$, $v = v(s)$. Since these planes depend on one parameter, they envelope a developable surface, and each has a characteristic line, the ruling of the developable which lies in it. As a matter of fact, the characteristic line of the tangent plane at the point P of C is the line through P in the direction conjugate to that of the tangent line to C at P. In other words:

THEOREM 1. *The characteristic lines of the tangent planes to the surface at the points of a curve C are the lines through the points of C in the directions which are conjugate to those of the tangent lines to C.*

The equation of the tangent plane to S at an arbitrary point P of C is

$$f \equiv (X - x\,|\,\zeta) = 0,$$

where $u = u(s)$, $v = v(s)$. The characteristic line of the tangent plane is the line in which it is intersected by the plane

$$f_s \equiv \left(X - x\,\middle|\,\frac{d\zeta}{ds}\right) = 0.$$

Hence, the characteristic line goes through P and is the line through P in the tangent plane which is perpendicular to the vector $d\zeta/ds$ at P, or, more simply, to the vector $d\zeta$. Thus, if $\delta v/\delta u$ is the direction of the characteristic line, then δx is perpendicular to $d\zeta$: $(\delta x | d\zeta) = 0$. Since

$$-(\delta x | d\zeta) = edu\delta u + f(du\delta v + dv\delta u) + gdv\delta v,$$

$(\delta x | d\zeta) = 0$ reduces, when the lines of curvature are parametric, to (27), and so the theorem is proved.

Furthermore, we have obtained the criterion for conjugate directions when the parametric curves are arbitrary.

THEOREM 2. *The two directions dv/du and $\delta v/\delta u$ at the elliptic or hyperbolic point P are conjugate directions if and only if at P*

$$edu\delta u + f(du\delta v + dv\delta u) + gdv\delta v = 0. \tag{28}$$

Conjugate systems. A system of curves on a surface is known as a conjugate system if the directions of the curves of the system at each and every elliptic or hyperbolic point are conjugate directions. The system of curves consisting of the two families of curves in the directions dv/du and $\delta v/\delta u$ is a conjugate system when and only when at every elliptic or hyperbolic point

$$edu\delta u + f(du\delta v + dv\delta u) + gdv\delta v = 0. \tag{29}$$

This equation is identically satisfied by $dv = 0$ and $\delta u = 0$ if and only if $f = 0$. In other words:

THEOREM 3. *A necessary and sufficient condition that the parametric curves form a conjugate system is that $f = 0$.*

Since the parametric curves form an orthogonal system if and only if $F = 0$, and the vanishing of F and f characterize the parametric curves as lines of curvature, we conclude that *a conjugate system is orthogonal when and only when it consists of lines of curvature.*

It is evident from (29) that the theory of conjugate systems bears the same relation to the second fundamental form of the surface as does the theory of orthogonal systems to the first fundamental form. Hence, we may state, without detailed proofs, the following theorems analogous to Theorems 2 and 3 of § 35.

THEOREM 4. *The system of curves defined by*

$$l du^2 + 2\, m du dv + n dv^2 = 0$$

is a conjugate system if and only if

$$en - 2\, fm + gl = 0.$$

THEOREM 5. *If a family of curves is given, there exists, in general, a second family of curves such that the two families form a conjugate system. If the differential equation of the given family is* $M du + N dv = 0$, *that of the second family is*

$$(eN - fM) du + (fN - gM) dv = 0.$$

The exception to the theorem will be discussed in the next paragraph.

EXERCISES

1. Prove that the parametric curves on the surface

$$x_1 = \cos u, \quad x_2 = \sin u + \sin v, \quad x_3 = \cos v$$

are all circles of radius unity and form a conjugate system.

2. Show that the parabolas in which the planes $x_1 =$ const. and $x_2 =$ const. meet the elliptic paraboloid $x_1^2/a^2 + x_2^2/b^2 = 2\, x_3$ form a conjugate system.

3. Find the family of curves on the right conoid which is conjugate to the orthogonal trajectories of the rulings.

4. Are the normal curvatures in two conjugate directions at a point ever equal?

5. Show that the sum of the radii of normal curvature at a point in two conjugate directions is the same for each two conjugate directions.

46. Asymptotic lines. *Imaginary points and directions.* Thus far, we have recognized only the real points of the surface $x = x(u, v)$. Henceforth, we shall frequently find it convenient to introduce imaginary points represented by pairs of values of u, v, at least one of which is imaginary. Similarly, we shall recognize in the tangent plane at a point imaginary, as well as real, directions.

In dealing with imaginary elements we agree to apply the same terminology as in the case of real elements, and, what is more important, we agree to extend to imaginary elements the formulas developed for real elements, so long as they retain meaning.

Asymptotic directions. If e, f, g are not all zero at a point P, the equation

$$edu^2 + 2\,fdudv + gdv^2 = 0 \tag{30}$$

defines at P two directions, which are real and distinct, real and coincident, or conjugate-imaginary, according as $d^2 < 0$, $d^2 = 0$, or $d^2 > 0$ at P. In other words:

THEOREM 1. *At a planar point, every direction is an asymptotic direction. At a nonplanar point, there exist just two asymptotic directions, which are real and distinct, real and coincident, or conjugate-imaginary, according as the point is hyperbolic, parabolic, or elliptic.*

The asymptotic directions at a point P were defined as the directions in which the tangent plane at P has contact of at least the second order. They may also be characterized as the directions at P in which the normal curvature is zero. At an elliptic or hyperbolic point, they are the directions of the asymptotes of the Dupin indicatrix, and, at a parabolic point, they coincide in a principal direction.

Asymptotic lines. A curve on the surface whose direction at each and every point is an asymptotic direction is an asymptotic line. In other words, the asymptotic lines are the curves defined by the differential equation (30).

THEOREM 2. *Every curve in a plane is an asymptotic line. On a surface, other than a plane, there exist two families of asymptotic lines. On a surface of negative curvature, these two families are real and distinct; on a surface of positive curvature, they are conjugate-imaginary; and, on a developable surface, they are real and coincident.*

The single (doubly counting) family of asymptotic lines on a developable surface consists of the rulings. For, the normal curvature at a point in the direction of the ruling which passes through the point is zero, and hence the single asymptotic direction at the point is that of the ruling. Since this direction is a principal direction, the rulings form also one of the families of lines of curvature. The second family consists of the orthogonal trajectories of the rulings and therefore, in the case of the tangent surface of a twisted curve, of the involutes of the curve.

It is a fact, with which the reader may be familiar, that there exist on every nonsingular quadric surface two families of straight lines. On a hyperboloid of one sheet or a hyperbolic paraboloid, these lines are real, whereas on an ellipsoid, a hyperboloid of two sheets, or an elliptic paraboloid, they are conjugate-imaginary. In every case, they constitute the asymptotic lines of the surface.

THEOREM 3. *A straight line on a surface is always an asymptotic line.*

For, the tangent planes at the points of the line contain the line, and so the direction of the line is always an asymptotic direction.

THEOREM 4. *A curve, not a straight line, is an asymptotic line if and only if the osculating plane at each point is the tangent plane to the surface at the point.*

The theorem follows from the formula,

$$\frac{\cos\phi}{R} = \frac{edu^2 + 2\,fdudv + gdv^2}{Edu^2 + 2\,Fdudv + Gdv^2},$$

for the curvature $1/R$ of the curve. Since $1/R \neq 0$, the curve is an integral curve of (30) when and only when $\phi = \pi/2$.

We give a second proof. If the given curve C is an asymptotic line, its tangential directions are conjugate to themselves. Hence, the characteristic lines of the tangent planes to the surface, S, at the points of C are the tangent lines to C. Thus, C is the edge of regression of the developable surface enveloped by the tangent planes. But the planes enveloping the tangent surface of a curve are the osculating planes of the curve. Consequently, the tangent planes to S at the points of C are the osculating planes of C. Conversely, if the osculating planes of C are the tangent planes to S, the characteristic lines of these tangent planes are the tangent lines to C, the tangential directions of C are self-conjugate and therefore asymptotic, and C is an asymptotic line.

Since asymptotic directions are self-conjugate, a family of asymptotic lines is conjugate to itself. Herein lies the exception to Theorem 5 of § 45.

THEOREM 5. *The asymptotic lines form an orthogonal system if and only if the surface is a minimal surface.*

For, the condition that the curves (30) form an orthogonal system, namely $Eg - 2\,Ff + Ge = 0$, is precisely the condition that $K' = 0$; see, also, § 42, Theorem 3.

THEOREM 6. *The parametric curves on a surface, other than a developable or a plane, are the asymptotic lines if and only if* $e = 0$, $g = 0$.

For, equation (30) is identical with the differential equation of the parametric curves when and only when $e = g = 0$.

Example. Since the circular helices on the right helicoid (16) have the rulings of the helicoid as their common principal normals, the osculating planes of the helices are the tangent planes to the helicoid. Hence, the helices form one of the two families of asymptotic lines on the helicoid. Inasmuch as the second family necessarily consists of the rulings, the asymptotic lines form an orthogonal system and the helicoid is a minimal surface. The same results may be obtained analytically by use of the data given in § 42.

EXERCISES

1. Find the finite equation of the second family of asymptotic lines on the right conoid.

2. Show that the asymptotic lines on the surface $x_3 = x_1^4 - x_2^4$ are the curves in which the surface is met by the two families of cylinders $x_1^2 + x_2^2 =$ const. and $x_1^2 - x_2^2 =$ const.

3. Show that the locus of points on the hyperbolic paraboloid $x_1^2/a^2 - x_2^2/b^2 = 2\,x_3$ at which the rulings cut at right angles is the hyperbola in which the paraboloid is cut by the plane $2\,x_3 = b^2 - a^2$. How many rulings of the one family cut a given ruling of the other family orthogonally?

Suggestion. Begin by finding a parametric representation for which the rulings are the parametric curves.

4. The surface $x_3 = f(x_1, x_2)$ has the representation $x_1 = x_1$, $x_2 = x_2$, $x_3 = f(x_1, x_2)$ in terms of x_1, x_2 as parameters. Find, in terms of the quantities

$$p = \frac{\partial f}{\partial x_1}, \quad q = \frac{\partial f}{\partial x_2}, \quad r = \frac{\partial^2 f}{\partial x_1^2}, \quad s = \frac{\partial^2 f}{\partial x_1 \partial x_2}, \quad t = \frac{\partial^2 f}{\partial x_2^2},$$

the fundamental forms and curvatures, the differential equations of the important systems of curves, and the conditions that the surface be minimal or developable.

47. Isometric systems. *Isotropic curves.* The simplest approach to the so-called isometric systems of curves on a surface is through the introduction as parametric curves of certain imaginary curves on the surface known as isotropic or minimal curves.

A curve is said to be isotropic when the distance, measured along the curve, between each two points of the curve is zero. In other words, a curve is isotropic if and only if $ds = 0$.

On the surface S: $x = x(u, v)$, there exist two conjugate-imaginary families of isotropic curves, defined by

$$ds^2 \equiv E du^2 + 2\,F du dv + G dv^2 = 0. \tag{31}$$

In order to introduce these curves as the parametric curves, we first factor the quadratic form:

$$ds^2 \equiv \left(\sqrt{E}\,du + \frac{F + iD}{\sqrt{E}}\,dv\right)\left(\sqrt{E}\,du + \frac{F - iD}{\sqrt{E}}\,dv\right),$$

and then introduce integrating factors for the two component expressions. Since these two expressions are conjugate-imaginary, it follows that, if $\sigma_1(u, v) + i\sigma_2(u, v)$, where σ_1, σ_2 are real functions of u, v, is an integrating factor of the first expression, then $\sigma_1(u, v) - i\sigma_2(u, v)$ is an integrating factor of the second. Hence, there exist two functions $\mu(u, v)$, $\nu(u, v)$ such that

$$\begin{aligned}(\sigma_1 + i\sigma_2)\left(\sqrt{E}\,du + \frac{F + iD}{\sqrt{E}}\,dv\right) &= d\mu,\\ (\sigma_1 - i\sigma_2)\left(\sqrt{E}\,du + \frac{F - iD}{\sqrt{E}}\,dv\right) &= d\nu.\end{aligned} \tag{32}$$

The finite equations of the two families of isotropic curves are, then, $\mu(u, v) = \text{const.}$ and $\nu(u, v) = \text{const.}$ In other words, the change of coordinates $\mu = \mu(u, v)$, $\nu = \nu(u, v)$ introduces the isotropic curves as the parametric curves. The linear element in (31) thereby takes on the form

$$ds^2 = \lambda d\mu d\nu, \tag{33}$$

where $\lambda = 1/(\sigma_1^2 + \sigma_2^2)$.

Isometric systems referred to isometric parameters. It is evident from (32) that μ and ν are conjugate-imaginary functions of u and v, that is, that they are of the form

$$\mu = u_1(u, v) + iv_1(u, v), \quad \nu = u_1(u, v) - iv_1(u, v),$$

where $u_1(u, v)$ and $v_1(u, v)$ are real functions of u, v. Hence $d\mu = du_1 + i dv_1$, $d\nu = du_1 - i dv_1$, and (33) becomes

$$ds^2 = \lambda(du_1^2 + dv_1^2). \tag{34}$$

The new parameters $u_1 = u_1(u, v)$, $v_1 = v_1(u, v)$ which we have thus introduced are real parameters. Hence, λ should be a real function of them. In fact, since $\lambda = 1/(\sigma_1^2 + \sigma_2^2)$ is a real function of u, v and u, v are real functions of u_1, v_1, λ is a real function of u_1, v_1.

Since the new parameters u_1, v_1 are real functions of u, v, the new parametric curves are real curves. The system which they form is known as an isometric, or isothermic, system, and the parameters u_1, v_1 are called isometric, or isothermic, parameters.

DEFINITION. *If a system of real curves can be introduced as parametric curves so that the linear element of the surface takes the form*

$$ds^2 = \lambda(du^2 + dv^2), \tag{35}$$

the system is said to be an isometric, or isothermic, system, and the parameters u, v are called isometric, or isothermic, parameters.

We have proved that there exists at least one isometric system on a given surface. We proceed now to determine all the isometric systems. To this end we assume that the given parameters u, v are isometric, and seek conditions necessary and sufficient that new parameters

$$\bar{u} = \bar{u}(u, v), \quad \bar{v} = \bar{v}(u, v) \tag{36}$$

be isometric. Since $x_u = x_{\bar{u}}\bar{u}_u + x_{\bar{v}}\bar{v}_u$, $x_v = x_{\bar{u}}\bar{u}_v + x_{\bar{v}}\bar{v}_v$, we have

$$\begin{aligned} E &= \bar{E}\bar{u}_u^2 + 2\,\bar{F}\bar{u}_u\bar{v}_u + \bar{G}\bar{v}_u^2, \\ F &= \bar{E}\bar{u}_u\bar{u}_v + \bar{F}(\bar{u}_u\bar{v}_v + \bar{u}_v\bar{v}_u) + \bar{G}\bar{v}_u\bar{v}_v, \\ G &= \bar{E}\bar{u}_v^2 + 2\,\bar{F}\bar{u}_v\bar{v}_v + \bar{G}\bar{v}_v^2. \end{aligned}$$

By hypothesis, $F = 0$ and $E = G$. Hence, the conditions that $\bar{u}$, $\bar{v}$ be isometric parameters, namely, $\bar{F} = 0$ and $\bar{E} = \bar{G}$, become

$$\begin{aligned} \bar{u}_u^2 + \bar{v}_u^2 &= \bar{u}_v^2 + \bar{v}_v^2, \\ \bar{u}_u\bar{u}_v + \bar{v}_u\bar{v}_v &= 0. \end{aligned} \tag{37}$$

If, for the moment, we interpret $\bar{u}_u$, $\bar{v}_u$ and $\bar{u}_v$, $\bar{v}_v$ respectively as the components of two vectors in a Cartesian plane, the second of these equations says that the two vectors are perpendicular, and the first, that they are equal in length. Consequently, if the vectors have their initial points at the origin, the second vector has, relative to the first, one of the two positions shown in Fig. 25. But, when the initial point of a vector is at the origin, the components of the vector are the coordinates of the terminal point. Hence, we conclude from the figure that equations (37) have the two solutions

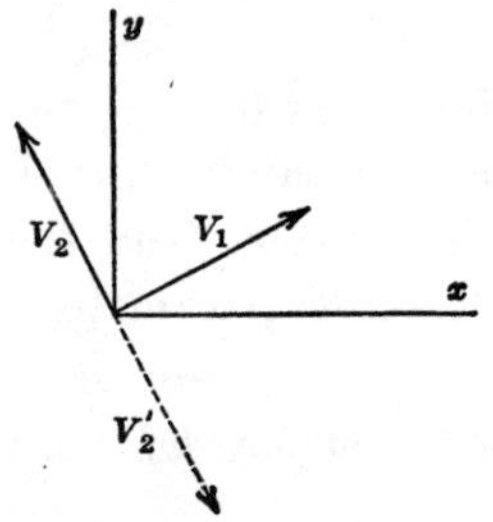

FIG. 25

$$\text{(38a)} \qquad \bar{u}_u = \bar{v}_v, \quad \bar{u}_v = -\bar{v}_u,$$

$$\text{(38b)} \qquad \bar{v}_u = \bar{u}_v, \quad \bar{v}_v = -\bar{u}_u.$$

Two functions $\bar{u}(u, v)$, $\bar{v}(u, v)$, in the order given, are said to be *conjugate functions* if they satisfy equations (38a).

Equations (38b) are precisely equations (38a) with the roles of $\bar{u}$ and $\bar{v}$ reversed. Hence, equations (38b) say that $\bar{v}$, $\bar{u}$ are conjugate functions. We may, then, state our results as follows:

THEOREM 1. *If u and v are isometric parameters, $\bar{u} = \bar{u}(u, v)$ and $\bar{v} = \bar{v}(u, v)$ are isometric parameters when and only when $\bar{u}(u, v)$ and $\bar{v}(u, v)$ are, in one or the other order, conjugate functions of u and v.*

If $\bar{u}$, $\bar{v}$ are, in the order given, conjugate functions, then $\bar{u} + i\bar{v}$ is a so-called analytic function of the complex variable $u + iv$, and conversely. Similarly, $\bar{v}$, $\bar{u}$ are conjugate functions if and only if $\bar{v} + i\bar{u}$ is an analytic function of $u + iv$.

Since there are infinitely many analytic functions of a complex variable, there are infinitely many pairs of isometric parameters, and infinitely many isometric systems on a given surface.

For example, $\bar{u} + i\bar{v} = (u + iv)^2$ is an analytic function of $u + iv$. Hence, if u and v are isometric parameters, then $\bar{u} = u^2 - v^2$, $\bar{v} = 2\,uv$ are isometric parameters, and the curves $u^2 - v^2 = \text{const.}$ and $2\,uv = \text{const.}$ form an isometric system.

Isometric systems referred to nonisometric parameters. Consider the change of parameters

$$\text{(39)} \qquad u = u(\bar{u}), \quad v = v(\bar{v}),$$

where u is an arbitrary function of $\bar{u}$, and v is an arbitrary function of $\bar{v}$. Evidently, when $u = \text{const.}$, $\bar{u} = \text{const.}$ and, when $v = \text{const.}$, $\bar{v} = \text{const.}$, and conversely. Hence, the new parametric curves $\bar{u} =$ const. and $\bar{v} =$ const. are identical with the original parametric curves $u =$ const. and $v =$ const. Thus, equations (39) represent a change of parameters which does not change the parametric curves. The same can be said of the equations $u = u(\bar{v})$, $v = v(\bar{u})$, but these equations are reducible to (39) by renaming the variables. In fact, every change of parameters which does not affect the parametric curves may be written in the form (39).

If, in the linear element (35) of a surface referred to isometric parameters, we make the change of parameters (39), the result is

$$ds^2 = \lambda(u'^2 d\bar{u}^2 + v'^2 d\bar{v}^2), \tag{40}$$

where $u' = du/d\bar{u}$, $v' = dv/d\bar{v}$. Hence, though the parametric curves, since they remain unaltered, still form an isometric system, the new parameters $\bar{u}$, $\bar{v}$ are not, in general, isometric parameters.

The linear element (40) is of the form

$$ds^2 = \bar{E}d\bar{u}^2 + \bar{G}d\bar{v}^2,$$

where $\bar{E} : \bar{G} = U(\bar{u}) : V(\bar{v})$. Conversely, if the linear element of a surface is of this form, the parametric curves form an isometric system. For, by hypothesis, $\bar{E} = \lambda U(\bar{u})$, $\bar{G} = \lambda V(\bar{v})$, where $U(\bar{u})$, $V(\bar{v})$ may both be assumed positive. Hence, we may write

$$ds^2 = \lambda(U(\bar{u})d\bar{u}^2 + V(\bar{v})d\bar{v}^2).$$

The change of parameters defined by $du = \sqrt{U(\bar{u})}\, d\bar{u}$, $dv = \sqrt{V(\bar{v})}\, d\bar{v}$ then reduces the linear element to the form (35).

The proposition thus established may be stated as follows:

THEOREM 2. *A necessary and sufficient condition that the parametric curves form an isometric system is that* $F = 0$ *and* $E : G = U(u) : V(v)$.

Example. In § 46, we showed that the parametric curves on the right helicoid, $x_1 = \bar{u} \cos \bar{v}$, $x_2 = \bar{u} \sin \bar{v}$, $x_3 = a\bar{v}$, are the asymptotic lines (the rulings and the circular helices), and in § 42, we found, as the linear element:

$$ds^2 = d\bar{u}^2 + (\bar{u}^2 + a^2)d\bar{v}^2 = (\bar{u}^2 + a^2)\left(\frac{d\bar{u}^2}{\bar{u}^2 + a^2} + d\bar{v}^2\right).$$

It follows, then, by Theorem 2, that the asymptotic lines form an isometric system.

In order to reduce the linear element to the form (35), we set $du = d\bar{u}/\sqrt{\bar{u}^2 + a^2}$, $dv = d\bar{v}$. Integrating and taking both the constants of integration equal to zero, we get $u = \sinh^{-1}(\bar{u}/a)$, $v = \bar{v}$ or $\bar{u} = a \sinh u$, $\bar{v} = v$. The linear element now becomes $ds^2 = a^2 \cosh^2 u(du^2 + dv^2)$, and the parameters u and v are isometric.

The equations, in terms of $\bar{u}$, $\bar{v}$, of the lines of curvature are $\sinh^{-1}(\bar{u}/a) - \bar{v} = \text{const.}$ and $\sinh^{-1}(\bar{u}/a) + \bar{v} = \text{const.}$ See § 42. The corresponding equations in u, v are $u - v = \text{const.}$, $u + v = \text{const.}$ Accordingly, when we make the change of parameters $u - v = u_1$, $u + v = v_1$, the new parametric curves, $u_1 = \text{const.}$ and $v_1 = \text{const.}$, are the lines of curvature. But u, v are isometric parameters and $u - v$ and $u + v$ are conjugate functions of u, v. Hence, the parameters u_1, v_1 are isometric and the lines of curvature form an isometric system.

EXERCISES

1. The parametric curves form an isometric system if and only if $F = 0$ and $\partial^2 \log (E/G)/\partial u \partial v = 0$.

2. Show that the curves which bisect the angles between the curves of an isometric system also form an isometric system.

3. Show that the meridians and parallels on the sphere of § 25 form an isometric system, that $u_1 = \log \tan \frac{1}{2}\phi$, $v_1 = \theta$ are isometric parameters, and that the parametric representation and linear element in terms of these parameters are $x_1 = a \operatorname{sech} u_1 \cos v_1$, $x_2 = a \operatorname{sech} u_1 \sin v_1$, $x_3 = -a \tanh u_1$ and $ds^2 = a^2 \operatorname{sech}^2 u_1(du_1^2 + dv_1^2)$.

4. Show that the rulings on a right conoid and their orthogonal trajectories form an isometric system if and only if the conoid is a right helicoid.

5. The lines of curvature on any quadric surface form an isometric system. Establish this fact in the case of a hyperbolic paraboloid.

6. Prove that the lines of curvature on a cylinder or a cone always form an isometric system, whereas those on the tangent surface of a twisted curve never do.

CHAPTER VI

THE FUNDAMENTAL THEOREM

48. The formulas of Gauss. The geometric properties of the surface $x = x(u, v)$ which we have thus far developed have been expressible in terms of the two fundamental differential quadratic forms and their coefficients, E, F, G, e, f, g. Is this true of all the properties of the surface? In other words: Is the surface determined, except for position, by the two fundamental forms?

If this question is to be answered in the affirmative, as we have every right to expect, a second question then presents itself. If two differential quadratic forms are given, will there exist a surface which has them as its two fundamental forms? Here, too, we should expect an affirmative answer, unless the coefficients E, F, G, e, f, g of the fundamental forms of a given surface are connected by relations which are identically satisfied. If this is the case, the coefficients of the given forms must be subjected to these relations before the given forms can possibly serve as the two fundamental forms of a surface.

As a matter of fact, there actually exist relations between E, F, G, e, f, g, and their partial derivatives. Preparatory to deducing them, we shall establish formulas of Gauss for x_{uu}, x_{uv}, x_{vv}.

According to § 5, an arbitrary vector δ can be written as a linear combination of three given vectors α, β, γ which are not parallel to a plane:

$$(\alpha\,\beta\,\gamma)\delta = (\delta\,\beta\,\gamma)\alpha + (\delta\,\gamma\,\alpha)\beta + (\delta\,\alpha\,\beta)\gamma. \tag{1}$$

It follows that any vector at an arbitrary point P on the given surface is expressible as a linear combination of three vectors at P which do not lie in a plane. As these three vectors it is natural to take x_u, x_v, ζ. Accordingly, we proceed to express x_{uu}, x_{uv}, x_{vv} as linear combinations of x_u, x_v, ζ.

Since $(x_u\,x_v\,\zeta) = D$, (1) becomes, for $\alpha = x_u$, $\beta = x_v$, $\gamma = \zeta$,

$$\delta = \frac{(\delta\,x_v\,\zeta)}{D}x_u + \frac{(\delta\,\zeta\,x_u)}{D}x_v + \frac{(\delta\,x_u\,x_v)}{D}\zeta,$$

or

$$\delta = \frac{(\delta\, x_v \,|\, x_u\, x_v)}{D^2}\, x_u + \frac{(x_u\, \delta \,|\, x_u\, x_v)}{D^2}\, x_v + (\delta \,|\, \zeta)\zeta.$$

Substituting x_{uu}, x_{uv}, x_{vv} in turn for δ, we get

$$\begin{aligned} x_{uu} &= C^1_{11}x_u + C^2_{11}x_v + e\zeta, \\ x_{uv} &= C^1_{12}x_u + C^2_{12}x_v + f\zeta, \\ x_{vv} &= C^1_{22}x_u + C^2_{22}x_v + g\zeta, \end{aligned} \tag{2}$$

where

$$\begin{aligned} C^1_{11} &= \frac{(x_{uu}\, x_v \,|\, x_u\, x_v)}{D^2}, & C^2_{11} &= \frac{(x_u\, x_{uu} \,|\, x_u\, x_v)}{D^2}, \\ C^1_{12} &= \frac{(x_{uv}\, x_v \,|\, x_u\, x_v)}{D^2}, & C^2_{12} &= \frac{(x_u\, x_{uv} \,|\, x_u\, x_v)}{D^2}, \\ C^1_{22} &= \frac{(x_{vv}\, x_v \,|\, x_u\, x_v)}{D^2}, & C^2_{22} &= \frac{(x_u\, x_{vv} \,|\, x_u\, x_v)}{D^2}. \end{aligned} \tag{3}$$

Formulas (2) are due to Gauss. The coefficients of x_u and x_v, which are here denoted by the use of the single letter C with two subscripts and one superscript, are frequently represented, especially in classical treatises, by the symbols

$$\begin{Bmatrix} 11 \\ 1 \end{Bmatrix}, \begin{Bmatrix} 12 \\ 1 \end{Bmatrix}, \begin{Bmatrix} 22 \\ 1 \end{Bmatrix}, \begin{Bmatrix} 11 \\ 2 \end{Bmatrix}, \begin{Bmatrix} 12 \\ 2 \end{Bmatrix}, \begin{Bmatrix} 22 \\ 2 \end{Bmatrix},$$

which were introduced by Christoffel. Accordingly, we shall refer to the C's as *Christoffel symbols*.

In expanding the composite inner products on the right-hand sides of (3), we need values for the scalar products of x_{uu}, x_{uv}, x_{vv} with x_u and x_v. These six scalar products appear in the six equations obtained by differentiating partially each of the equations $E = (x_u \,|\, x_u)$, $F = (x_u \,|\, x_v)$, $G = (x_v \,|\, x_v)$, first with respect to u and then with respect to v. Moreover, the six equations can be readily solved for the six scalar products and yield for them the values:

$$\begin{aligned} (x_{uu} \,|\, x_u) &= \tfrac{1}{2}E_u, & (x_{uu} \,|\, x_v) &= F_u - \tfrac{1}{2}E_v, \\ (x_{uv} \,|\, x_u) &= \tfrac{1}{2}E_v, & (x_{uv} \,|\, x_v) &= \tfrac{1}{2}G_u, \\ (x_{vv} \,|\, x_u) &= F_v - \tfrac{1}{2}G_u, & (x_{vv} \,|\, x_v) &= \tfrac{1}{2}G_v. \end{aligned} \tag{4}$$

Expansion of the right-hand sides of (3) now gives us

$$
(5)\quad
\begin{aligned}
C_{11}^1 &= \frac{GE_u - F(2F_u - E_v)}{2D^2}, & C_{11}^2 &= \frac{E(2F_u - E_v) - FE_u}{2D^2},\\
C_{12}^1 &= \frac{GE_v - FG_u}{2D^2}, & C_{12}^2 &= \frac{EG_u - FE_v}{2D^2},\\
C_{22}^1 &= \frac{G(2F_v - G_u) - FG_v}{2D^2}, & C_{22}^2 &= \frac{EG_v - F(2F_v - G_u)}{2D^2}.
\end{aligned}
$$

Thus, the Christoffel symbols are expressible in terms of E, F, G and their first partial derivatives.

Besides the formulas of Gauss, we shall need, for the purpose in view, expressions for ζ_u and ζ_v. Since $(\zeta_u | \zeta) = 0$ and $(\zeta_v | \zeta) = 0$, the vectors ζ_u and ζ_v lie in the tangent plane at P and hence are expressible as linear combinations of x_u and x_v. In fact,

$$
(6)\quad
\begin{aligned}
\zeta_u &= \frac{Ff - Ge}{D^2} x_u + \frac{Fe - Ef}{D^2} x_v,\\
\zeta_v &= \frac{Fg - Gf}{D^2} x_u + \frac{Ff - Eg}{D^2} x_v.
\end{aligned}
$$

These formulas may be established by the method employed to obtain those of Gauss. The method of undetermined coefficients is, however, just as simple in this case. We have, for example,

$$
(7)\qquad \zeta_u = A_1 x_u + B_1 x_v.
$$

Taking the inner product of each side of this equation, first with x_u and then with x_v, we obtain the equations

$$
EA_1 + FB_1 = -e, \qquad FA_1 + GB_1 = -f.
$$

When these equations are solved for A_1 and B_1 and the values found are substituted in (7), the result is the first of the equations (6). The second equation may be similarly established.

49. The equations of Gauss and Codazzi. By reasoning which is now familiar to the reader, x_{uuu}, x_{uuv}, x_{uvv}, x_{vvv} are expressible as linear combinations of x_u, x_v, ζ. These linear combinations are most easily obtained by differentiating the formulas of Gauss, and then replacing the derivatives x_{uu}, x_{uv}, x_{vv}, ζ_u, ζ_v which appear in the right-hand members of the resulting equations by their values in terms of x_u, x_v, ζ, as given by formulas (2) and (6). For example, when we differentiate the first of the Gauss formulas with respect to u, we get

$$x_{uuu} = C_{11}^1 x_{uu} + C_{11}^2 x_{uv} + e\zeta_u + \frac{\partial C_{11}^1}{\partial u} x_u + \frac{\partial C_{11}^2}{\partial u} x_v + e_u \zeta,$$

and we then reduce the expression on the right to a linear combination of x_u, x_v, ζ by replacing x_{uu}, x_{uv}, and ζ_u by their values in terms of x_u, x_v, ζ.

There are two ways of applying this process to get x_{uuv}; we may differentiate x_{uu} with respect to v or x_{uv} with respect to u. Similarly, x_{uvv} may be obtained by differentiating x_{uv} with respect to v or x_{vv} with respect to u. In each case, the two linear combinations of x_u, x_v, ζ obtained must be equal, or, what is the same thing, the difference between them must be identically zero. Hence, we have

$$\frac{\partial x_{uu}}{\partial v} - \frac{\partial x_{uv}}{\partial u} \equiv a_1 x_u + b_1 x_v + c_1 \zeta = 0,$$

$$\frac{\partial x_{uv}}{\partial v} - \frac{\partial x_{vv}}{\partial u} \equiv a_2 x_u + b_2 x_v + c_2 \zeta = 0.$$

Since the vectors x_u, x_v, ζ are not parallel to a plane, it follows that

$$a_1 = 0, \quad b_1 = 0, \quad c_1 = 0, \qquad a_2 = 0, \quad b_2 = 0, \quad c_2 = 0. \tag{8}$$

Actual computation shows that no one of the expressions a_1, a_2, b_1, b_2, c_1, c_2 reduces formally to zero. Hence, the six equations (8) constitute relations between E, F, G, e, f, g, and their partial derivatives.

It so happens that the first four of these equations, namely, $a_1 = 0$, $a_2 = 0$, $b_1 = 0$, $b_2 = 0$, all reduce to a single equation. This equation is due to Gauss and is known as *the Gauss equation.* It may be written in either of the following forms:

$$K = \frac{1}{D}\left[\frac{\partial}{\partial v}\left(\frac{D}{E} C_{11}^2\right) - \frac{\partial}{\partial u}\left(\frac{D}{E} C_{12}^2\right)\right], \tag{9a}$$

$$K = \frac{1}{D}\left[\frac{\partial}{\partial u}\left(\frac{D}{G} C_{22}^1\right) - \frac{\partial}{\partial v}\left(\frac{D}{G} C_{12}^1\right)\right], \tag{9b}$$

where $K = d^2/D^2$ is the total curvature of the surface. It is to be noted that these two forms are symmetric, one to the other; that is, that each becomes the other when u and v, and hence E and G, and the indices 1 and 2, are interchanged.

The last two equations of (8), namely, $c_1 = 0$ and $c_2 = 0$, are, when written out,

$$
\begin{aligned}
&e_v - f_u - C^1_{12}e + (C^1_{11} - C^2_{12})f + C^2_{11}g = 0, \\
&g_u - f_v + C^1_{22}e + (C^2_{22} - C^1_{12})f - C^2_{12}g = 0.
\end{aligned} \tag{10}
$$

These equations are symmetric to one another, but, unlike equations (9a) and (9b), they are not equivalent. They are generally known as *the Codazzi equations*, inasmuch as Codazzi discovered equations equivalent to them. As a matter of fact, prior to Codazzi, Mainardi had made similar discoveries, and equations (10), themselves, though not given in the work of Gauss, are readily deducible from his results.

Equations (10), taken with one of the equations (9), are commonly known as *the Gauss-Codazzi equations*. These three equations constitute the only relations between E, F, G, e, f, g and their derivatives, in that every other relation can be derived from them by the processes of algebra and the calculus. For example, it is conceivable that the identity

$$\frac{\partial \zeta_u}{\partial v} - \frac{\partial \zeta_v}{\partial u} \equiv a_3 x_u + b_3 x_v + c_3 \zeta = 0,$$

which is obtained from the values of ζ_u and ζ_v in (6) by the process described above, might yield new relations: $a_3 = 0$, $b_3 = 0$, $c_3 = 0$. Actually, c_3 reduces formally to zero, and the equations $a_3 = 0$ and $b_3 = 0$ are those of Codazzi.

We are now ready to answer the questions propounded in § 48.

FUNDAMENTAL THEOREM. *If E, F, G, e, f, g are given functions of u, v which satisfy the Gauss-Codazzi equations, there exists a surface, $x = x(u, v)$, uniquely determined except for its position in space, which has respectively as its first and second fundamental forms the quadratic forms $Edu^2 + 2\,Fdudv + Gdv^2$ and $edu^2 + 2\,fdudv + gdv^2$, provided merely that the first of these quadratic forms is positive definite, that is, that $EG - F^2 > 0$ and $E > 0$, $G > 0$.*

The theorem was first proved by Bonnet in 1867. It answers both the questions of § 48. It says, not only that a surface exists subject to the given conditions, but also that if two surfaces have the same fundamental forms, the surfaces are congruent. In other words, all the properties of a surface which are independent of its position in space are expressible in terms of its two fundamental forms and their coefficients.

Let us look more closely at the Gauss-Codazzi equations. The equations of Codazzi are linear and homogeneous in e, f, g and their first partial derivatives, with coefficients C which involve E, F, G and their first partial derivatives.

The right-hand member in either form of the Gauss equation, since it involves derivatives of the C's, is a function of $E, F, G,$ and their first and second partial derivatives. The quantities e, f, g enter into the equation only through the total curvature K. In other words:

THEOREM 2. *The total curvature of a surface is expressible in terms of the coefficients E, F, G of the first fundamental form and their first and second partial derivatives.*

This remarkable theorem is perhaps the most important in all surface theory. Its geometrical significance will become more apparent later. Analytically, it says that the total curvature of a surface is completely determined by the first fundamental form. Accordingly, we may speak of the curvature determined by a quadratic form, or, more simply, of the *curvature of a quadratic form.*

In bringing this discussion to a close, we note a form of the Gauss equation which is due to Frobenius:

$$-4D^4K = 2D^3\left(\frac{\partial}{\partial u}\frac{G_u - F_v}{D} - \frac{\partial}{\partial v}\frac{F_u - E_v}{D}\right) + \begin{vmatrix} E & F & G \\ E_u & F_u & G_u \\ E_v & F_v & G_v \end{vmatrix}.$$

Similar forms of the Codazzi equations were found by Study.*

EXERCISES

1. Show that

$$\frac{\partial \log D}{\partial u} = C^1_{11} + C^2_{12}, \qquad \frac{\partial \log D}{\partial v} = C^1_{12} + C^2_{22}$$

and hence that

$$\frac{\partial \omega}{\partial u} = -\frac{D}{E}C^2_{11} - \frac{D}{G}C^1_{12}, \qquad \frac{\partial \omega}{\partial v} = -\frac{D}{E}C^2_{12} - \frac{D}{G}C^1_{22},$$

* See Blaschke, *Differentialgeometrie*, Vol. 1, third edition, p. 117.

where ω is the angle between the parametric curves. By means of the latter identities establish the equivalence of the two expressions (9) for K.

2. Prove that, if the curvature of the quadratic form $ds^2 = \lambda(u)(du^2 + dv^2)$ is unity, $\lambda(u) = c^2 \operatorname{sech}^2 (cu + d)$, $c \neq 0$, and that the form can then be reduced to $ds^2 = \operatorname{sech}^2 u_1(du_1^2 + dv_1^2)$; see § 47, Ex. 3.

3. The foregoing problem, if the curvature of the quadratic form is -1.

50. Spherical representation. In addition to the fundamental forms $(dx|dx)$ and $-(dx|d\zeta)$, we shall now introduce a third fundamental form, namely, the linear element, $(d\zeta|d\zeta)$, of the spherical representation of the surface.

The spherical representation, $\zeta = \zeta(u, v)$, is two-dimensional only when the surface is an ordinary surface. In this case, we shall restrict the surface, if necessary, so that the normal vector ζ has, in general, a different oriented direction at each point. There will be, then, in general, a one-to-one correspondence between the points of the surface and the points of the spherical representation, and hence a one-to-one correspondence between the points of the spherical representation and the permissible pairs of values of u, v.

The linear element of the spherical representation is

$$(11) \qquad (d\zeta|d\zeta) = \mathcal{E}du^2 + 2\,\mathcal{F}dudv + \mathcal{G}dv^2,$$

where

$$(12) \qquad \mathcal{E} = (\zeta_u|\zeta_u), \qquad \mathcal{F} = (\zeta_u|\zeta_v), \qquad \mathcal{G} = (\zeta_v|\zeta_v).$$

Its discriminant is

$$(13) \qquad \mathfrak{D}^2 = \mathcal{E}\mathcal{G} - \mathcal{F}^2 = (\zeta_u\,\zeta_v|\zeta_u\,\zeta_v).$$

Evidently, $\mathfrak{D} = \sqrt{\zeta_u\,\zeta_v|\zeta_u\,\zeta_v}$ is non-negative when the given surface is ordinary, and is identically zero when the given surface is a developable or a plane.

The third fundamental form and its coefficients and discriminant behave in the same way, with respect to rigid motions and changes of parameters, as do the first fundamental form and its coefficients and discriminant; see § 39.

Since the first and second fundamental forms determine the surface completely, the third fundamental form must be expressible in terms of them. To find the actual relationship, we assume that the lines of curvature are parametric. Then the equation $d\zeta = -dx/r$ of Olinde Rodrigues holds for the parametric curves, and we have

$$\zeta_u = -x_u/r_1, \qquad \zeta_v = -x_v/r_2, \tag{14}$$

where $1/r_1$, $1/r_2$ are the principal normal curvatures. Hence

$$\begin{aligned} (dx\,|\,dx) &= Edu^2 + Gdv^2, \\ -(dx\,|\,d\zeta) &= \frac{1}{r_1}Edu^2 + \frac{1}{r_2}Gdv^2, \\ (d\zeta\,|\,d\zeta) &= \frac{1}{r_1^2}Edu^2 + \frac{1}{r_2^2}Gdv^2. \end{aligned} \tag{15}$$

Eliminating Edu^2 and Gdv^2, we obtain the equation

$$\begin{vmatrix} (dx\,|\,dx) & 1 & 1 \\ -(dx\,|\,d\zeta) & \frac{1}{r_1} & \frac{1}{r_2} \\ (d\zeta\,|\,d\zeta) & \frac{1}{r_1^2} & \frac{1}{r_2^2} \end{vmatrix} = 0,$$

which, when we expand the determinant and divide by $1/r_1 - 1/r_2$, becomes

$$K(dx\,|\,dx) + K'(dx\,|\,d\zeta) + (d\zeta\,|\,d\zeta) = 0. \tag{16}$$

At an umbilic, $1/r_1 = 1/r_2$ and we cannot divide by $1/r_1 - 1/r_2$. The relation (16) is, however, still valid. For, if $1/a$ is the constant value of $1/r$ at the umbilic, then $K = 1/a^2$, $K' = 2/a$, and $d\zeta = -dx/a$ for every direction of departure. Hence, (16) holds.

We have established relation (16) under the hypothesis that the lines of curvature are parametric. But the separate terms in (16) are invariant with respect to a change of parameters. Consequently, the relation holds when the parametric curves are arbitrarily chosen.

Equivalent to (16), when the parametric curves are arbitrary, are the equations

$$\begin{aligned} KE - K'e + \mathcal{E} &= 0, \\ KF - K'f + \mathcal{F} &= 0, \\ KG - K'g + \mathcal{G} &= 0. \end{aligned} \tag{17}$$

We next note that

$$\mathfrak{D}^2 = K^2 D^2.$$

This relation is true when the lines of curvature are parametric, as is evident from (15). That it is always valid follows, then, from the fact that D^2 and $\mathfrak{D}^2$ are relative invariants, and K^2 is an absolute invariant, with respect to a change of parameters. From it we conclude that

$$\mathfrak{D} = \pm\, KD, \tag{18}$$

where the plus sign or the minus sign is to be taken according as $K > 0$ or $K < 0$.

Since $d^2 = KD^2$ and $KD = \pm\, \mathfrak{D}$, it follows that $d^2 = \pm\, D\mathfrak{D}$. Comparison of this equation with equation (40) of Chapter IV, namely, $d^2 = D(\zeta\; \zeta_u\; \zeta_v)$, gives us $\mathfrak{D} = \pm\, (\zeta\; \zeta_u\; \zeta_v)$. Hence, since $D = (\zeta\; x_u\; x_v)$, we have

$$D = (\zeta\; x_u\; x_v), \quad d^2 = (\zeta\; x_u\; x_v)(\zeta\; \zeta_u\; \zeta_v), \quad \mathfrak{D} = \pm\, (\zeta\; \zeta_u\; \zeta_v). \tag{19}$$

Finally, we note

$$\widehat{x_u\; x_v} = D\zeta, \qquad \widehat{\zeta_u\; \zeta_v} = \pm\, \mathfrak{D}\zeta. \tag{20}$$

The first of these formulas is familiar; the second follows from equation (35) of Chapter IV, namely, $\widehat{\zeta_u\; \zeta_v} = (\zeta\; \zeta_u\; \zeta_v)\zeta$.

Properties of the spherical representation. If a point P of the surface is a parabolic or planar point, it is evident from (18) that the corresponding point P' of the sphere fails to be regular. Accordingly, in discussing the representation of the surface on the sphere, we assume that the surface is ordinary and consider only the elliptic and hyperbolic points on it.

The first of equations (20) says that the direction of rotation about the point P on the surface from the positive direction of the u-curve to the positive direction of the v-curve, as viewed from the terminal point of the normal vector, ζ, at P, is counterclockwise (§ 36). The corresponding direction of rotation, about the point P' on the sphere which corresponds to P, is that from the positive direction of the u-curve at P' to the positive direction of the v-curve. We agree to view this direction of rotation about P' from outside the sphere, for example, from the terminal point of the normal vector, ζ, to the sphere at P'. The second equation in (20) then tells us that the direction of rotation about P' is counterclockwise or clockwise according as $K > 0$ or $K < 0$ at P. Hence, *corresponding directions of rotation about a point P of the surface and the corresponding point P' of the spherical representation are the same or opposite according as P is an elliptic or a hyperbolic point of the surface.*

It follows from (17) or (15) that, when $F = 0$ and $f = 0$, then $\mathcal{F} = 0$, that is, that if the parametric curves on the surface are lines of curvature, the parametric curves on the sphere form an orthogonal system. Thus, *a system of lines of curvature on the surface is represented by an orthogonal system of curves on the sphere.*

Equations (14) say that *a line of curvature on the surface and the curve representing it on the sphere have parallel tangents at corresponding points.* The lines of curvature are the only curves with this property. For, the vector dx/ds tangent to the curve $x = x(u(s), v(s))$ on the surface is parallel always to the vector $d\zeta/ds$ tangent to the curve $\zeta = \zeta(u(s), v(s))$ on the sphere if and only if $1/r$ exists so that $d\zeta + dx/r = 0$ is an identity in s. But the curve on the surface is then a line of curvature.

Surface determined by the second and third fundamental forms. It is evident from (17) that, if the surface is ordinary, E, F, G are expressible in terms of $e, f, g, \mathcal{E}, \mathcal{F}, \mathcal{G}$, and K, K'. But K, K' may be expressed in terms of $e, f, g, \mathcal{E}, \mathcal{F}, \mathcal{G}$. We have, in fact,

$$(21) \qquad K = \frac{\mathfrak{D}^2}{d^2}, \qquad K' = \frac{\mathcal{E}g - 2\,\mathcal{F}f + \mathcal{G}e}{d^2}.$$

Thus, E, F, G are expressible in terms of $e, f, g, \mathcal{E}, \mathcal{F}, \mathcal{G}$. Hence, an ordinary surface is uniquely determined by its second and third fundamental forms.

The first of equations (21) follows from (19), since it is evident from (19) that $d^2/D^2 = \mathfrak{D}^2/d^2$. To establish the second equation, we note that the usual expression for K' (§ 41) may be rewritten in the form $K' = (KEg - 2\,KFf + KGe)/d^2$. Substituting for KE, KF, KG the values given for them by (17), we obtain an equation which reduces to the one desired.

EXERCISES

1. Show that the system of curves on the sphere which represents the asymptotic lines on a right helicoid is an isometric orthogonal system.

2. The same for the system of curves on the sphere which represents the lines of curvature on the torus.

3. Prove that corresponding to an asymptotic direction at a point of an ordinary surface is the direction at the corresponding point of the spherical representation which is perpendicular to the asymptotic direction.

4. Show that the angles between the asymptotic lines on a surface of negative curvature are supplements of the corresponding angles on the spherical representation.

5. If two directions at a point of an ordinary surface are conjugate directions, each is perpendicular to the spherical representation of the other.

6. The angles between the curves of a conjugate system at a point of a surface are equal or supplementary to the corresponding angles on the spherical representation according as the total curvature of the surface at the point is negative or positive.

CHAPTER VII

GEODESIC CURVATURE. GEODESICS

51. Geodesic curvature. Let there be given on the surface S: $x = x(u, v)$ a directed curve C: $u = u(s)$, $v = v(s)$. Let P: (u, v) be an arbitrary point of C, and let $\overline{C}$ be the projection of C on the tangent plane to S at P. The curvature at P of $\overline{C}$, taken with a sign which we shall prescribe presently, is known as the geodesic curvature of C at P, and shall be denoted by $1/\rho$.

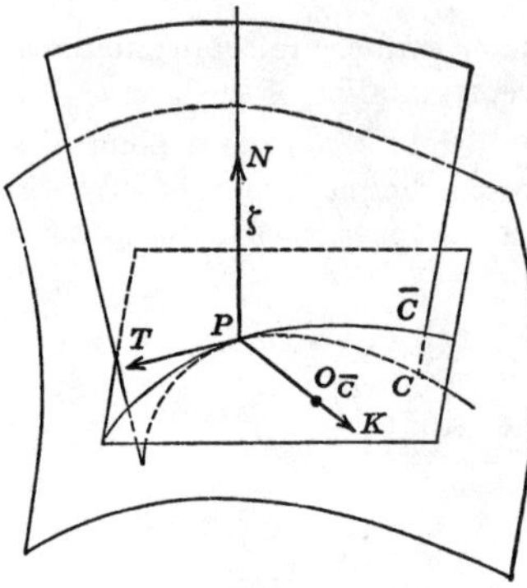

FIG. 26

Unless $1/\rho = 0$, when there is no call for a sign, the center of curvature, $O_{\overline{C}}$, of $\overline{C}$ exists and lies on the principal normal PK of $\overline{C}$ at P. We direct PK so that the directed tangent PT, the directed line PK, and the directed normal to the surface have the same disposition as the axes. We then take $\rho = \overline{PO_{\overline{C}}}$. In other words, the geodesic curvature of C at P is the curvature of $\overline{C}$ at P or the negative thereof, according as the directed line-segment $\overline{PO_{\overline{C}}}$ has the same sense as the directed line PK or the opposite sense. If the positive direction on C is reversed, the geodesic curvature changes sign; for, the positive direction of the tangent to C is then reversed, and hence so is the sense of the line PK.

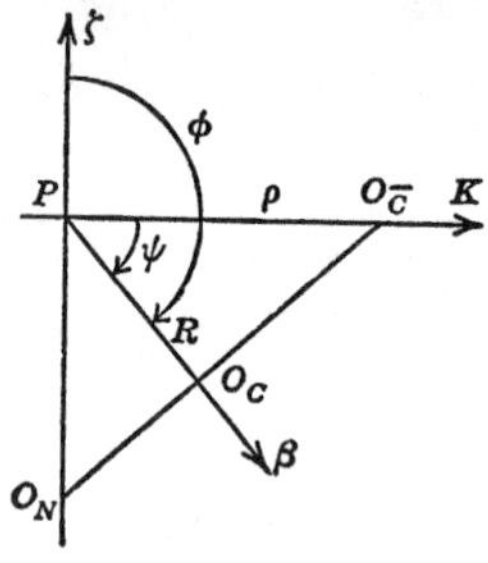

FIG. 27

The curve $\overline{C}$ is the section by the tangent plane at P of the cylinder formed by the lines through the points of C perpendicular to the tangent plane. More specifically, it is the normal section of the cylinder in the direction of the curve C. Hence, by Meusnier's Theorem, applied to the cylinder, the center of curvature, O_C, of C is the projection, on the principal normal to C, of the center of geodesic curvature, $O_{\overline{C}}$ (Fig. 27). But, by Meusnier's Theorem applied to S, O_C is the projection, on the principal normal of C, of the center of normal curvature, O_N, of S in the direction of C. Consequently, $O_{\overline{C}}$ is the point in which the line joining O_N and O_C meets the line PK.

To express this result analytically, we introduce, in the plane of PK and ζ, a positive direction of rotation for the measurement of angle, namely, the direction which is clockwise when the plane is viewed from the terminal point of the tangent vector to C at P. Then, if ψ is the angle from the directed line PK to the principal normal vector β, we have $\rho \cos \psi = R$, where $1/R$ is the curvature of C at P. Hence,

$$\frac{1}{\rho} = \frac{\cos \psi}{R}. \tag{1}$$

Furthermore, if ϕ is the angle from ζ to β, $\cos \psi = \sin \phi$, and

$$\frac{1}{\rho} = \frac{\sin \phi}{R}. \tag{2}$$

Thus, the geodesic curvature is positive or negative according as the directed ray PK and the vector β lie on the same side of the normal to the surface or on opposite sides of this normal.

THEOREM 1. *The geodesic curvature of a curve is equal numerically to the ordinary curvature at every point of the curve if and only if the curve is an asymptotic line.*

For, equation (1) says that the numerical value of $1/\rho$ is equal to $1/R$ for a curve, not a straight line, if and only if the osculating plane of the curve at each point is the tangent plane to the surface at the point. But, the curve is then an asymptotic line, and conversely (§ 46). On the other hand, a straight line is always an asymptotic line, and both curvatures of it are zero.

Beltrami's formula for geodesic curvature. Since the unit vector at P in the direction of PK is $\widehat{\zeta\,\alpha}$, where α is the tangent vector to C at P, $\cos \psi = (\widehat{\zeta\,\alpha} \mid \beta)$. But $\alpha = dx/ds$ and $\beta/R = d^2x/ds^2$. Hence

$$\frac{\cos \psi}{R} = \left(\zeta \; \frac{dx}{ds} \; \frac{d^2x}{ds^2}\right),$$

and we obtain for the geodesic curvature of C the expression

$$(3) \qquad \frac{1}{\rho} = \frac{1}{D}\left(x_u\, x_v \middle| \frac{dx}{ds}\, \frac{d^2x}{ds^2} \right).$$

Now

$$(4) \qquad \frac{dx}{ds} = x_u \frac{du}{ds} + x_v \frac{dv}{ds},$$

$$(5) \qquad \frac{d^2x}{ds^2} = x_u \frac{d^2u}{ds^2} + x_v \frac{d^2v}{ds^2} + x_{uu}\left(\frac{du}{ds}\right)^2 + 2\, x_{uv} \frac{du}{ds}\frac{dv}{ds} + x_{vv}\left(\frac{dv}{ds}\right)^2.$$

Substituting these values in (3), and expanding, we obtain Beltrami's formula,

$$(6) \qquad \frac{1}{\rho} = D\left[\frac{du}{ds}\frac{d^2v}{ds^2} - \frac{dv}{ds}\frac{d^2u}{ds^2} + C^2_{11}\left(\frac{du}{ds}\right)^3 + (2\,C^2_{12} - C^1_{11})\left(\frac{du}{ds}\right)^2 \frac{dv}{ds} + (C^2_{22} - 2\,C^1_{12})\frac{du}{ds}\left(\frac{dv}{ds}\right)^2 - C^1_{22}\left(\frac{dv}{ds}\right)^3 \right],$$

where the C's are the Christoffel symbols (3) of Chapter VI.

Recalling equations (5) of Chapter VI, we arrive at the following conclusion.

THEOREM 2. *The geodesic curvature of a curve may be expressed in terms of E, F, G, and their first partial derivatives, in conjunction with the functions defining the curve.*

In other words, geodesic curvature, unlike ordinary curvature or normal curvature, depends only on the first fundamental form of the surface.

If the given curve is a u-curve, $dx/ds = x_u/\sqrt{E}$ and therefore, by (4), $du/ds = 1/\sqrt{E}$ and $dv/ds = 0$. On the other hand, for a v-curve, $du/ds = 0$ and $dv/ds = 1/\sqrt{G}$. Hence, we obtain, as the geodesic curvatures, $1/\rho_1$ and $1/\rho_2$, of the u-curves and v-curves,

$$(7) \qquad \frac{1}{\rho_1} = C^2_{11}\frac{D}{E^{3/2}}, \qquad \frac{1}{\rho_2} = -\,C^1_{22}\frac{D}{G^{3/2}}.$$

If the parametric curves form an orthogonal system, then $F = 0$ and the values of C^2_{11} and C^1_{22}, as given by Chapter VI, (5), become $-E_v/(2\,G)$ and $-G_u/(2\,E)$. Thus, we find, in this case,

$$(8) \qquad \frac{1}{\rho_1} = -\frac{1}{\sqrt{G}}\frac{\partial \log \sqrt{E}}{\partial v}, \qquad \frac{1}{\rho_2} = \frac{1}{\sqrt{E}}\frac{\partial \log \sqrt{G}}{\partial u}.$$

EXERCISES

1. Show that the square of the curvature of a curve on a surface at a point is equal to the sum of the squares of the geodesic curvature of the curve at the point and the normal curvature at the point in the direction of the curve: $(1/R)^2 = (1/\rho)^2 + (1/r)^2$.

2. If the curves of one family of an isometric system of curves on a surface have constant geodesic curvature, so also have the curves of the other. Prove this theorem and show that a surface possesses an isometric system of curves of this character if and only if its linear element can be put into the form

$$ds^2 = \frac{du^2 + dv^2}{(U(u) + (V(v))^2}.$$

52. Geodesics. A curve, not a straight line, whose principal normal at every point coincides with the normal to the surface at the point, is known as a geodesic on the surface. A straight line is always to be reckoned as a geodesic.

The meridians on a surface of revolution, that is, the various positions of the plane curve whose revolution about the axis generates the surface, are geodesics on the surface, as is immediately verified by application of the definition. In particular, every great circle on a sphere is a geodesic on the sphere.

Since the principal normal of a plane curve, if it exists, lies always in the plane of the curve, it follows that the geodesics in a plane are the straight lines.

THEOREM 1. *A curve on a developable surface, other than a straight line, is a geodesic if and only if the surface is the rectifying developable of the curve.*

According to the definition of the rectifying developable of a curve, not a straight line, the tangent plane to the developable at a point of the curve is the rectifying plane of the curve at the point. Hence, the normal to the developable coincides always with the principal normal to the curve, and the curve is a geodesic on the developable. Conversely, if a curve, not a straight line, is a geodesic on a developable surface, the tangent planes to the surface are the rectifying planes of the curve and therefore the surface is the rectifying developable of the curve.

The following theorem is a direct consequence of (2), for a curve which is not a straight line, and is obvious for a straight line.

THEOREM 2. *A curve is a geodesic on a surface if and only if its geodesic curvature is identically zero.*

The theorem says that characteristic of a geodesic is that its projection on the tangent plane at an arbitrary point behaves at the point like a straight line.

From (7), we conclude the following proposition.

THEOREM 3. *The u-curves on a surface are geodesics if and only if $C^2_{11} = 0$; and, the v-curves, if and only if $C^1_{22} = 0$.*

It is worth while to give a geometric proof. The vanishing of C^2_{11}, for example, is equivalent to the vanishing of $(x_u\, x_v \,|\, x_u\, x_{uu})$ or $(\zeta \,|\, \widehat{x_u\, x_{uu}})$. But $\widehat{x_u\, x_{uu}}$ is a vector in the direction of the binormal of the general u-curve, and $(\zeta \,|\, \widehat{x_u\, x_{uu}}) = 0$ is, therefore, a condition necessary and sufficient that the u-curve be a geodesic. The argument breaks down if the u-curve is a straight line; but, then, on the one hand, $\widehat{x_u\, x_{uu}} = 0$, and, on the other, a straight line is a geodesic.

The next theorem is a direct consequence of (8).

THEOREM 4. *If the parametric curves form an orthogonal system, the u-curves are geodesics when and only when E is a function of u alone, and the v-curves are geodesics when and only when G is a function of v alone.*

If both the u-curves and the v-curves are geodesics, E and G are respectively functions, $U(u)$ and $V(v)$, of u and v. Hence, since $F = 0$, the linear element is

$$ds^2 = U(u)du^2 + V(v)dv^2,$$

and may be reduced, by virtue of the change of parameters defined by $d\bar{u} = \sqrt{U}\, du$, $d\bar{v} = \sqrt{V}\, dv$, to

$$ds^2 = d\bar{u}^2 + d\bar{v}^2.$$

The total curvature, K, of this quadratic form is zero, as is at once evident from the Frobenius form of the Gauss equation given in § 49. Thus, we arrive at the following conclusion.

THEOREM 5. *If there exists on a surface an orthogonal system of geodesics, the surface is a developable or a plane.*

EXERCISES

1. Show that the straight lines on a surface are the only asymptotic lines which are geodesics.

2. When is a particular parallel on a surface of revolution a geodesic?

3. Prove that the evolutes of a curve are geodesics on the polar developable.

4. A geodesic, not a straight line, is a plane curve if and only if it is a line of curvature.

5. If a line of curvature is a geodesic, it is a plane curve. Is the converse true?

6. A necessary and sufficient condition that the curvature of a line of curvature be equal, within sign, to the corresponding principal normal curvature is that the line of curvature be a geodesic.

7. A line of curvature on an ordinary surface is a geodesic if and only if its spherical representation is a great circle, or a portion thereof.

8. Show that, if two families of geodesics on a surface cut under a constant angle, the surface is a developable or a plane.

53. Geodesic parallels. The orthogonal trajectories of a family of lines in the plane have the property that the segments cut from the lines by any two of them are all equal. For, if the lines of the family are all parallel or all go through a point, the fact is obvious. Otherwise, the lines envelope a curve, their orthogonal trajectories are the involutes of this curve, and we know that each two involutes of a curve cut equal segments from the tangents to the curve,—here the lines of the given family.

The orthogonal trajectories of a family of geodesics on an arbitrary surface have the same property: each two of them cut equal segments from the geodesics. Conversely, if each two orthogonal trajectories of a family of curves on a surface cut equal segments from the curves of the family, the curves of the family are geodesics on the surface. In other words:

THEOREM 1. *A necessary and sufficient condition that the segments cut from the curves of a family of curves on a surface by two arbitrarily chosen orthogonal trajectories of the family be all equal is that the curves of the family be geodesics on the surface.*

Let the curves of the given family be the u-curves and let their orthogonal trajectories be the v-curves. Then, since the differential of arc of the u-curve, $v = v_0$, is $ds = \sqrt{E(u, v_0)}\, du$, the segment cut from this u-curve by the v-curves, $u = u_1$ and $u = u_2$ $(u_2 > u_1)$, is given by the integral

$$s = \int_{u_1}^{u_2} \sqrt{E(u, v_0)}\, du.$$

This segment is the same for every u-curve if and only if the integral is independent of v_0, and it can be shown that this is the case if and only if E is a function of u alone. But, since the parametric curves form an orthogonal system, the u-curves are geodesics when and only when E does not involve v. Hence, the theorem is proved.

The orthogonal trajectories of a family of geodesics are known as *geodesic parallels*. They are called parallels, since each two of them are equally distant, and, geodesic parallels, since the distances in question are measured along geodesics.

An example of geodesic parallels is to be had in the so-called parallels on a surface of revolution, that is, the curves traced by the individual points of the plane curve which is rotated to generate the surface. The parallels of latitude on a sphere are, then, geodesic parallels.

A family of geodesic parallels can consist of geodesics, according to § 52, Theorem 5, only if the surface is a developable or a plane. Even in these cases, geodesic parallels are not, in general, geodesics; parallel curves in the plane, for example, are not, in general, straight lines.

Geodesic parameters. According to the previous discussion, a necessary and sufficient condition that the u-curves be geodesics and the v-curves be the geodesic parallels orthogonal to them is that $E = U(u)$ and $F = 0$, that is, that the linear element be of the form

$$ds^2 = U(u)du^2 + Gdv^2.$$

But it is then possible, by setting $d\bar{u} = \sqrt{U(u)}\, du$, $\bar{v} = v$ and afterward dropping the bars, to reduce the linear element to the form

$$ds^2 = du^2 + Gdv^2. \tag{9}$$

Inasmuch as E is now unity, the distance along an arbitrary u-curve from the geodesic parallel $u = 0$ to the geodesic parallel

$u = u$ is equal precisely to u. Hence, the parameter u is now the common arc of all the geodesics, measured from one of the geodesic parallels.

Conversely, if the u-curves are geodesics and have the parameter u as their common arc, and the v-curves are the geodesic parallels orthogonal to them, then the linear element is necessarily of the form (9). For, since the element of arc of an arbitrary u-curve, namely, $\sqrt{E}\,du$, reduces to du, it follows that $E = 1$. But $F = 0$ by hypothesis, and hence the statement is proved.

The result thus established may be stated as follows:

THEOREM 2. *A necessary and sufficient condition that the linear element of a surface be of the form* (9) *is that the u-curves be geodesics with the parameter u as their common arc, and the v-curves be the geodesic parallels orthogonal to them.*

The parameters u and v are then called *geodesic parameters.* On account of the simplicity of the form (9), these parameters are frequently the most convenient to which to refer the surface. When they are employed, the Gauss equation, for example, becomes simply

$$K = -\frac{1}{\sqrt{G}}\frac{\partial^2\sqrt{G}}{\partial u^2}. \tag{10}$$

Geodesics as curves of shortest distance. We remind the reader that by a family of curves on a surface we mean a one-parameter set of curves which has the property that through a given point of the surface there passes, in general, just one curve of the set. By a family of geodesics we mean, then, a one-parameter set of geodesics with this property. For our present purposes, we demand, further, that the property hold, not merely in general, but without exception. The family of geodesics is then known as a *field.*

It is to be remarked that a field of geodesics does not have to cover the whole surface. As a matter of fact, more often than not, the field covers only a portion of the surface. The only fields of geodesics which cover the entire plane, for example, are the families of parallel lines. Again, there is no field of geodesics which covers the whole sphere, inasmuch as each two great circles (geodesics) on the sphere intersect.

THEOREM 3. *If P_1 and P_2 are two points of a geodesic which can be imbedded in a field of geodesics, then the arc P_1P_2 of the geodesic is shorter than any other arc which connects P_1 and P_2 and lies entirely in the portion of the surface covered by the field.*

Let the portion of the surface in question be referred to geodesic parameters u, v, where the u-curves are the geodesics of the field. If, then, $v = v_0$ is the given geodesic, the points P_1 and P_2 have the coordinates (u_1, v_0) and (u_2, v_0). Hence, if we assume that $u_2 > u_1$, the geodesic arc P_1P_2 has the length $u_2 - u_1$.

An arbitrary curve which passes through P_1 and P_2 and lies within the prescribed region on the surface may be represented by an equation of the form $v = \phi(u)$, where $\phi(u_1) = v_0$ and $\phi(u_2) = v_0$. The arc P_1P_2 of this curve is given, according to (9), by the integral

$$s = \int_{u_1}^{u_2} \sqrt{1 + G(u, \phi(u))\phi'^2}\, du.$$

Now, $G > 0$, and, unless the curve is the given geodesic, $\phi' \not\equiv 0$. Hence, for a curve other than the geodesic, the integrand is, in general, greater than unity, and the integral is, therefore, greater than $u_2 - u_1$. Thus, the theorem is proved.

54. Differential equations of the geodesics. According as the curve $C: u = u(s),\ v = v(s)$ is, or is not, a straight line, the vector d^2x/ds^2 is a null vector or a vector in the direction of the principal normal to C. Hence C is a geodesic if and only if the (proper or null) vector d^2x/ds^2 is perpendicular to each of two nonparallel vectors in the tangent plane to the surface.

If we take as these vectors x_u and x_v, we obtain the pair of equations

$$(11) \qquad \left(x_u \middle| \frac{d^2x}{ds^2}\right) = 0, \qquad \left(x_v \middle| \frac{d^2x}{ds^2}\right) = 0,$$

characterizing the curve C as a geodesic.

Instead of x_u and x_v, we may take the vectors $\widehat{\zeta\, x_u}$ and $\widehat{\zeta\, x_v}$. We then get, after slight changes in form, the equations

$$(12) \qquad \left(x_u\, x_v \middle| \frac{d^2x}{ds^2}\, x_v\right) = 0, \qquad \left(x_u\, x_v \middle| x_u\, \frac{d^2x}{ds^2}\right) = 0.$$

When we set for d^2x/ds^2 its value, as given by (5), these equations become

$$(13)\qquad \begin{aligned} &\frac{d^2u}{ds^2} + C^1_{11}\left(\frac{du}{ds}\right)^2 + 2\,C^1_{12}\frac{du}{ds}\frac{dv}{ds} + C^1_{22}\left(\frac{dv}{ds}\right)^2 = 0, \\ &\frac{d^2v}{ds^2} + C^2_{11}\left(\frac{du}{ds}\right)^2 + 2\,C^2_{12}\frac{du}{ds}\frac{dv}{ds} + C^2_{22}\left(\frac{dv}{ds}\right)^2 = 0. \end{aligned}$$

On the other hand, if we should set for d^2x/ds^2 its value in equations (11), the terms in d^2u/ds^2 and d^2v/ds^2 in the resulting equations would be, respectively,

$$E\frac{d^2u}{ds^2} + F\frac{d^2v}{ds^2}, \qquad F\frac{d^2u}{ds^2} + G\frac{d^2v}{ds^2}.$$

These equations would not be so simple as equations (13), and the result of solving them simultaneously for d^2u/ds^2 and d^2v/ds^2 would be precisely equations (13). For, equations (11) and equations (12), since they both characterize C as a geodesic, are equivalent.

THEOREM 1. *The curve C represented parametrically in terms of the arc by the equations $u = u(s)$, $v = v(s)$ is a geodesic if and only if the functions $u(s)$, $v(s)$ satisfy the differential equations* (13), *or the equivalent equations* (11).

Equations (13) are two ordinary differential equations of the second order in the two dependent variables u, v and the one independent variable s. According to the theory of differential equations, they have a unique solution $u = u(s)$, $v = v(s)$ such that, for a given value s_0 of s, the functions $u(s)$, $v(s)$, du/ds, dv/ds have prescribed values, u_0, v_0, $(du/ds)_0$, $(dv/ds)_0$. Hence, there exists a unique geodesic, $u = u(s)$, $v = v(s)$, passing through a given point (u_0, v_0) and having at the point a prescribed direction $(dx/ds)_0$:

$$(14)\qquad \left(\frac{dx}{ds}\right)_0 = x_u(u_0, v_0)\left(\frac{du}{ds}\right)_0 + x_v(u_0, v_0)\left(\frac{dv}{ds}\right)_0.$$

THEOREM 2. *There is a unique geodesic passing through a given regular point on the surface and having at the point a prescribed direction.*

The proof of the theorem is not yet complete. The equations $u = u(s)$, $v = v(s)$ obtained as the solution of (13) certainly represent a curve which goes through (u_0, v_0) and has there the required direction $(dx/ds)_0$. But, in order to apply Theorem 1, so as to conclude that this curve is a geodesic, we must know that s is its arc, that is, that $(\alpha \mid \alpha) = 1$, where $\alpha = dx/ds$.

Since $u = u(s)$, $v = v(s)$ satisfy equations (11) as well as (13),

$$\frac{1}{2}\frac{d}{ds}\left(\frac{dx}{ds}\middle|\frac{dx}{ds}\right) \equiv \left(x_u\middle|\frac{d^2x}{ds^2}\right)\frac{du}{ds} + \left(x_v\middle|\frac{d^2x}{ds^2}\right)\frac{dv}{ds} = 0,$$

and hence

$$(15) \qquad \left(\frac{dx}{ds}\middle|\frac{dx}{ds}\right) = \text{const.}$$

We return now to (14) and make the natural demand that $(du/ds)_0$ and $(dv/ds)_0$ be so chosen that $(dx/ds)_0$ is a unit vector. Then the left-hand side of (15) is equal to unity for $s = s_0$ and hence identically equal to unity. The proof of the theorem is thus complete.

The theorem implies that there is a two-parameter family of geodesics on a given surface.

EXERCISES

1. Prove geometrically and analytically that, if one of the equations (11) is satisfied for a curve which is not a parametric curve, so is the other.

2. Show that a curve, other than a u-curve (v-curve), is a geodesic if and only if the first (second) of equations (11) is satisfied.

3. Find the finite equation of the geodesics on a surface whose linear element is

$$ds^2 = v(du^2 + dv^2), \qquad v > 0.$$

Solution. Since the u-curves are not geodesics, all the geodesics are defined, according to Ex. 2, by the first of equations (11). This equation becomes in the present case

$$v\frac{d^2u}{ds^2} + \frac{du}{ds}\frac{dv}{ds} = 0,$$

and hence has the first integral

$$v\frac{du}{ds} = k.$$

Eliminating ds from this relation and the equation defining the linear element, we obtain the differential equation

$$(v - k^2)du^2 - k^2dv^2 = 0,$$

whose solution is

$$(u - c)^2 = 4k^2(v - k^2).$$

4. Determine the geodesics on a surface whose linear element is

$$ds^2 = \lambda(u)(du^2 + dv^2).$$

5. Find the finite equation of the geodesics on the paraboloid of revolution, $x_1 = u \cos v$, $x_2 = u \sin v$, $2x_3 = au^2$.

6. A surface on which there exists a conjugate system of geodesics is known as a *surface of Voss*. Show that a right helicoid is a surface of Voss in an infinity of ways depending on one parameter.

Suggestion. Show that the geodesics on the helicoid of § 42 are the integral curves of the differential equation $k^2du^2 - (u^2 + a^2)(u^2 + a^2 - k^2)dv^2 = 0$, where k is an arbitrary constant.

7. Prove that the curvature of a twisted geodesic on a cone of revolution varies inversely as the cube of the distance from the vertex to the point tracing the geodesic.

55. Bonnet's formula for geodesic curvature. Let the curve C now be defined by an equation of the form

$$\phi(u, v) = \text{const.} \tag{16}$$

Then $\phi_u du + \phi_v dv = 0$ and hence $du/ds = k\phi_v$, $dv/ds = -k\phi_u$, where s is the arc of C. Since $Edu^2 + 2Fdudv + Gdv^2 = ds^2$, we find that $k = \pm 1/S$, where

$$S = \sqrt{E\phi_v^2 - 2F\phi_v\phi_u + G\phi_u^2}.$$

Consequently, $du/ds = \pm \phi_v/S$, $dv/ds = \mp \phi_u/S$ and the tangent vector α to C is

$$\alpha = \pm \frac{\phi_v x_u - \phi_u x_v}{\sqrt{E\phi_v^2 - 2F\phi_v\phi_u + G\phi_u^2}}, \tag{17}$$

where the sign depends on the direction on C in which s is measured.

According to (3), the geodesic curvature $1/\rho$ of the directed curve C is

$$\frac{1}{\rho} = \frac{1}{D}\left(x_u\, x_v \middle|\, \alpha \frac{d\alpha}{ds}\right). \tag{18}$$

Since α, as given by (17), is a function simply of u and v,

$$\frac{d\alpha}{ds} = \alpha_u \frac{du}{ds} + \alpha_v \frac{dv}{ds}.$$

Substituting this value of $d\alpha/ds$ in (18), we have

$$\frac{D}{\rho} = (\widehat{x_u\, x_v} | \alpha\, \alpha_u) \frac{du}{ds} + (\widehat{x_u\, x_v} | \alpha\, \alpha_v) \frac{dv}{ds}.$$

Since $\alpha = x_u(du/ds) + x_v(dv/ds)$,

$$\widehat{x_u\, x_v}\frac{du}{ds} = \widehat{\alpha\, x_v}, \qquad \widehat{x_u\, x_v}\frac{dv}{ds} = \widehat{x_u\, \alpha}.$$

Therefore

$$\frac{D}{\rho} = (\alpha\, x_v \,|\, \alpha\, \alpha_u) + (x_u\, \alpha \,|\, \alpha\, \alpha_v).$$

Inasmuch as $(\alpha \,|\, \alpha) = 1$, $(\alpha \,|\, \alpha_u) = 0$ and $(\alpha \,|\, \alpha_v) = 0$. Thus

$$\text{(19)} \qquad \frac{D}{\rho} = (x_v \,|\, \alpha_u) - (x_u \,|\, \alpha_v).$$

Since

$$\frac{\partial}{\partial u}(x_v \,|\, \alpha) = (x_v \,|\, \alpha_u) + (x_{uv} \,|\, \alpha), \quad \frac{\partial}{\partial v}(x_u \,|\, \alpha) = (x_u \,|\, \alpha_v) + (x_{uv} \,|\, \alpha),$$

it follows that

$$(x_v \,|\, \alpha_u) - (x_u \,|\, \alpha_v) = \frac{\partial}{\partial u}(x_v \,|\, \alpha) - \frac{\partial}{\partial v}(x_u \,|\, \alpha),$$

and we obtain, finally,

$$\text{(20)} \qquad \frac{1}{\rho} = \frac{1}{D}\left[\frac{\partial}{\partial u}(x_v \,|\, \alpha) - \frac{\partial}{\partial v}(x_u \,|\, \alpha)\right].$$

The result of substituting for α the expression given in (17) is Bonnet's formula for the geodesic curvature of the curve C, namely,

$$\text{(21)} \quad \frac{1}{\rho} = \pm\frac{1}{D}\left[\frac{\partial}{\partial u}\frac{F\phi_v - G\phi_u}{\sqrt{E\phi_v^2 - 2\,F\phi_v\phi_u + G\phi_u^2}} + \frac{\partial}{\partial v}\frac{F\phi_u - E\phi_v}{\sqrt{E\phi_v^2 - 2\,F\phi_v\phi_u + G\phi_u^2}}\right].$$

The sign depends on which direction along C is taken as the positive direction.

Bonnet's formula is simpler in structure and more convenient for most applications than that of Beltrami. On the other hand, Beltrami's formula presents no ambiguity.

EXERCISES

1. The curves C_1 and C_2 of an orthogonal system on a surface are so directed that at each point the directed angle from the directed curve C_1 to the directed curve C_2 is $\pi/2$. If $1/\rho_1$, $1/\rho_2$ are the geodesic curvatures, and ds_1 and ds_2 the differentials of arc, of the curves C_1 and C_2 respectively, a necessary and sufficient condition that the system of curves be isometric is that

$$\frac{d}{ds_1}\left(\frac{1}{\rho_1}\right) + \frac{d}{ds_2}\left(\frac{1}{\rho_2}\right) = 0.$$

Suggestion. Employ equations (8) and the theorem of § 47, Ex. 1.

2. *Liouville's formula for geodesic curvature.* With reference to an orthogonal system of directed curves on a surface such as described in the preceding exercise, the geodesic curvature of an arbitrary directed curve C: $u = u(s)$, $v = v(s)$ is given by the formula

$$\frac{1}{\rho} = \frac{\cos\theta}{\rho_1} + \frac{\sin\theta}{\rho_2} + \frac{d\theta}{ds},$$

where θ is the directed angle at an arbitrary point P of C from the directed curve C_1 through P to the directed curve C. Starting with equation (20), establish this formula.

3. A one-parameter set of families of curves on a surface which has the property that each two families of the set intersect under a constant angle is called a *pencil of families of curves.* Show that the sum of the squares of the geodesic curvatures of two orthogonal families of curves belonging to a given pencil is the same for each two orthogonal families of the pencil.

4. Prove that the expression on the left-hand side of the equation in Ex. 1 has the same value for each two orthogonal families of a pencil of families of curves. Hence, show that every pair of orthogonal families of the pencil forms an isometric system if one pair does.

5. The geodesic curvature of a line of curvature on an ordinary surface is equal, except perhaps for sign, to the geodesic curvature of its spherical representation multiplied by the corresponding principal normal curvature. In establishing this theorem, make use of equations (15) and (10) of Chapter VI.

6. The geodesic curvature of the spherical representation of a line of curvature is equal, except perhaps for sign, to the tangent of the angle which the principal normal of the line of curvature makes with the normal to the surface.

56. Geodesic torsion. *Torsion of a geodesic.* Let there be given on a surface a geodesic, not a straight line, and let α, β, γ be the vectors of the trihedral of the geodesic at an arbitrary point P. Then the torsion, $1/\tau$, of the geodesic at P, obtained from the formula $d\beta/ds = -\alpha/R - \gamma/\tau$ by taking the inner product of each member with γ, is $1/\tau = -(d\beta/ds\,|\,\gamma)$. But $\beta = \pm\zeta$ and $\gamma = \widehat{\alpha\,\beta}$, where $\alpha = dx/ds$. Hence,

$$\frac{1}{\tau} = -\left(\frac{d\zeta}{ds}\,\frac{dx}{ds}\,\zeta\right) = -\frac{(d\zeta\; dx\; \zeta)}{ds^2}. \tag{22}$$

Since $dx = x_u du + x_v dv$ and $d\zeta = \zeta_u du + \zeta_v dv$,

$$(d\zeta\, dx\, \zeta) = (\zeta_u\, x_u\, \zeta)du^2 + [(\zeta_u\, x_v\, \zeta) + (\zeta_v\, x_u\, \zeta)]dudv + (\zeta_v\, x_v\, \zeta)dv^2.$$

Setting $\zeta = \widehat{x_u\, x_v}/D$ and evaluating each of the resulting composite inner products, we obtain, finally, as the expanded form of (22),

$$(23)\qquad \frac{1}{\tau} = -\frac{(Ef - Fe)du^2 + (Eg - Ge)dudv + (Fg - Gf)dv^2}{D(Edu^2 + 2\,Fdudv + Gdv^2)}.$$

THEOREM 1. *A geodesic, not a straight line, is a plane curve if and only if it is a line of curvature.*

For, the result of equating to zero the right-hand member of (23) is precisely the differential equation of the lines of curvature.

Incidentally, since every curve on a sphere is a line of curvature, every geodesic on the sphere is a plane curve, a fact which we know to be true.

Geodesic torsion of the surface at a point in a given direction. The expression (23) is similar in structure to the normal curvature; it depends only on the point P and on the direction dv/du at P. We shall call it the geodesic torsion of the surface at P in the direction dv/du.

Though the torsion at P of the geodesic which issues from P in a given direction fails to exist when the geodesic is a straight line, the geodesic torsion always exists. For, (23) defines $1/\tau$ for each and every direction at P, regardless of the nature of the geodesic in the direction. Moreover, when this geodesic is a straight line, it is not in general true that the geodesic torsion is zero. For the geodesic torsion vanishes only in the principal directions, whereas the direction of a straight line is always an asymptotic direction.

We shall next deduce a formula for geodesic torsion analogous to Euler's formula for normal curvature. Assuming that the lines of curvature are parametric, and introducing the principal normal curvatures $1/r_1 = e/E$ and $1/r_2 = g/G$, we find that

$$\frac{1}{\tau} = -\frac{Eg - Ge}{\sqrt{EG}}\frac{du}{ds}\frac{dv}{ds} = \left(\frac{1}{r_1} - \frac{1}{r_2}\right)\sqrt{E}\frac{du}{ds}\sqrt{G}\frac{dv}{ds}.$$

If θ is the directed angle from the positive direction of the u-curve to the direction dv/du, we have, according to Chapter V, (14),

$$\cos\theta = \pm\sqrt{E}\frac{du}{ds}, \qquad \sin\theta = \pm\sqrt{G}\frac{dv}{ds}.$$

For the positive direction of the u-curve, $ds = \sqrt{E}\,du$ and $\theta = 0$; and, for the positive direction of the v-curve, $ds = \sqrt{G}\,dv$ and $\theta = \pi/2$. Accordingly, the plus sign must be taken in both formulas. Hence,

$$\frac{1}{\tau} = \frac{1}{2}\left(\frac{1}{r_1} - \frac{1}{r_2}\right)\sin 2\,\theta. \tag{24}$$

At an umbilic, the geodesic torsion is zero in every direction. Excluding this case for a moment, we conclude from (24) that the geodesic torsion is zero in the principal directions, and takes on its extreme values in the two (perpendicular) directions which bisect the angles between the principal directions.

Since $\sin 2(\theta + \pi/2) = -\sin 2\theta$, and $\sin 2(-\theta) = -\sin 2\theta$, we draw the following conclusions.

THEOREM 2. *The geodesic torsions at a point in two perpendicular directions, or in two directions which are equally inclined to a principal direction, are negatives of one another.*

In particular, the two extreme values of the geodesic torsion are negatives of one another, and the geodesic torsions in the two asymptotic directions are negatives of one another.

Geodesic torsion of a curve. By the geodesic torsion of a curve C at a point P is meant the geodesic torsion of the surface at P in the direction which C has at P. All the curves through P which have at P the same direction have, then, the same geodesic torsion at P.

The results which we have obtained for the geodesic torsion of the surface may be reworded to apply to curves. We note, for example, that the geodesic torsion of a curve is identically zero if and only if the curve is a line of curvature.

The geodesic torsion of the curve C: $u = u(s)$, $v = v(s)$ is given by either (22) or (23). The relation between it and the ordinary torsion, $1/T$, of C involves the directed angle ϕ from the vector ζ, normal to the surface, to the principal normal vector, β, to C. If the positive direction for the measurement of ϕ

is taken as that from the binormal vector γ to the vector β, then

$$\cos\phi = (\zeta|\beta), \qquad \sin\phi = (\zeta|\gamma).$$

Differentiating the first of these relations with respect to the arc s of C, and employing the second to simplify the result, we get

$$\sin\phi\left(\frac{d\phi}{ds} - \frac{1}{T}\right) + \left(\beta\left|\frac{d\zeta}{ds}\right.\right) = 0. \tag{25}$$

Now

$$\left(\beta\left|\frac{d\zeta}{ds}\right.\right) = \left(\widehat{\gamma\,\alpha}\left|\frac{d\zeta}{ds}\right.\right) = \left(\gamma\left|\widehat{\alpha\frac{d\zeta}{ds}}\right.\right).$$

Since $d\zeta/ds$ and $\alpha = dx/ds$ are both in the tangent plane, $\widehat{\alpha\, d\zeta}/ds = k\zeta$. Taking the inner product of each side of this relation with ζ, we find, by referring to (22), that $k = 1/\tau$. Hence,

$$\left(\gamma\left|\widehat{\alpha\frac{d\zeta}{ds}}\right.\right) = \frac{1}{\tau}(\zeta|\gamma) = \frac{\sin\phi}{\tau},$$

and (25) becomes

$$\sin\phi\left(\frac{d\phi}{ds} - \frac{1}{T} + \frac{1}{\tau}\right) = 0.$$

It follows that

$$\frac{1}{\tau} = \frac{1}{T} - \frac{d\phi}{ds}. \tag{26}$$

For, the conclusion is obvious if $\sin\phi \neq 0$. And, if $\sin\phi = 0$, C is a geodesic and hence, by definition, $1/\tau = 1/T$ and $d\phi/ds = 0$, so that (26) is automatically satisfied.

Relation (26) is due to Bonnet. It comprises the following theorem.

THEOREM 3. *If C is a curve, not a straight line, which passes through the point P and has at P a given direction, the difference, at P, of the torsion of C and the rate of change, with respect to the arc of C, of the angle which the osculating plane of C makes with the tangent plane to the surface is the same for all curves which pass through P and have at P the given direction, and is equal to the geodesic torsion at P in the given direction.*

Equation (26) also gives the condition that $1/T = 1/\tau$:

THEOREM 4. *The ordinary torsion and the geodesic torsion of a curve C, not a straight line, are identically equal if and only if the osculating plane of C makes a constant angle with the tangent plane to the surface.*

The curved geodesics form one case in point, and the curved asymptotic lines another.

It follows from (26) that, if $1/\tau = 0$, then $1/T = 0$ when and only when $d\phi/ds = 0$. In other words:

THEOREM 5. *A necessary and sufficient condition that a line of curvature, other than a straight line, be a plane curve is that its osculating plane make always the same angle with the tangent plane to the surface.*

Suppose, now, that C is the curve of intersection of two surfaces, S_1 and S_2, and that $1/\tau_1$ and $1/\tau_2$ are the values of $1/\tau$, and ϕ_1 and ϕ_2 are the values of ϕ, for the two surfaces. Then, since $1/T$ and s are independent of the surface on which C lies, we have, from (26),

$$\frac{1}{\tau_1} - \frac{1}{\tau_2} + \frac{d(\phi_1 - \phi_2)}{ds} = 0.$$

Hence, the angle $\phi_1 - \phi_2$ is constant when and only when $1/\tau_1$ and $1/\tau_2$ are equal. Thus:

THEOREM 6. *Two surfaces intersect under a constant angle if and only if the curve of intersection has the same geodesic torsion with respect to the one surface as it has with respect to the other.*

Joachimthal's theorem to the effect that, if the surfaces intersect under a constant angle, the curve of intersection is a line of curvature on both surfaces or on neither follows as a corollary. For, if $1/\tau_1$ and $1/\tau_2$ are the same, either both are zero, or neither is zero.

EXERCISES

1. The square of the geodesic torsion at a point in an asymptotic direction is equal to the negative of the total curvature at the point.

2. Prove that the foregoing theorem is also true of the square of the torsion of a curved asymptotic line at a point which is not an umbilic.

57. The trihedral of a curve on a surface. In the treatment of a space curve, considered by itself, the Frenet-Serret formulas, giving the rates of change with respect to the arc of the curve of the vectors of the trihedral α, β, γ proved of great value. We shall now develop similar formulas for a trihedral associated with a curve C considered as a curve on a surface.

To form the new trihedral at a point P of C, we take the unit vector α tangent to C at P, the unit vector ζ normal to the surface at P, and the unit vector $\nu = \widehat{\zeta\,\alpha}$ which is normal to C at P, lies in the tangent plane at P to the surface, and is so oriented that α, ν, ζ have, in the order given, the same disposition as the axes. The three vectors α, ν, ζ of the trihedral are, then, the unit vectors at P which have respectively the same directions and senses as the directed lines PT, PK, PN of Fig. 26.

Inasmuch as the vectors α, ν, ζ at P do not lie in a plane, any vector δ at P is expressible as a linear combination of them. In particular, since α, ν, ζ are mutually perpendicular unit vectors, it is readily shown, either by the method of undetermined coefficients or by use of the symbolic identity (18) of Chapter I, that

$$\delta = (\alpha\,|\,\delta)\alpha + (\nu\,|\,\delta)\nu + (\zeta\,|\,\delta)\zeta.$$

Substituting $d\alpha/ds$, $d\nu/ds$, $d\zeta/ds$ for δ, we obtain the preliminary forms,

$$\text{(27)}\qquad \begin{aligned} \frac{d\alpha}{ds} &= \left(\alpha\,\middle|\,\frac{d\alpha}{ds}\right)\alpha + \left(\nu\,\middle|\,\frac{d\alpha}{ds}\right)\nu + \left(\zeta\,\middle|\,\frac{d\alpha}{ds}\right)\zeta, \\ \frac{d\nu}{ds} &= \left(\alpha\,\middle|\,\frac{d\nu}{ds}\right)\alpha + \left(\nu\,\middle|\,\frac{d\nu}{ds}\right)\nu + \left(\zeta\,\middle|\,\frac{d\nu}{ds}\right)\zeta, \\ \frac{d\zeta}{ds} &= \left(\alpha\,\middle|\,\frac{d\zeta}{ds}\right)\alpha + \left(\nu\,\middle|\,\frac{d\zeta}{ds}\right)\nu + \left(\zeta\,\middle|\,\frac{d\zeta}{ds}\right)\zeta, \end{aligned}$$

of the formulas for the new trihedral which are analogous to those of Frenet-Serret.

The determinant of the coefficients of α, ν, ζ in equations (27) is skew-symmetric: each element in the principal diagonal is zero and each two elements symmetrically situated with respect to this diagonal are negatives of one another. This is simply another consequence of the fact that α, ν, ζ are mutually perpendicular unit vectors. From, say, $(\alpha\,|\,\alpha) = 1$ and $(\alpha\,|\,\nu) = 0$, we get, for example, $(\alpha\,|\,d\alpha/ds) = 0$ and $(\alpha\,|\,d\nu/ds) + (\nu\,|\,d\alpha/ds) = 0$. Hence, the conclusion is established.

It follows that, in order to evaluate all the coefficients in (27), we need to know only the values of $(\nu\,|\,d\alpha/ds)$, $(\nu\,|\,d\zeta/ds)$, and

$(\alpha|d\zeta/ds)$. These are respectively $1/\rho$, $1/\tau$, and $-1/r$, where $1/\rho$ and $1/\tau$ are the geodesic curvature and geodesic torsion of C at P and $-1/r$ is the normal curvature of the surface at P in the direction of C. For, according to (1),

$$\frac{1}{\rho} = \frac{(\nu|\beta)}{R} = \left(\nu\middle|\frac{\beta}{R}\right) = \left(\nu\middle|\frac{d\alpha}{ds}\right),$$

and, by (22),

$$\frac{1}{\tau} = -\left(\frac{d\zeta}{ds}\,\frac{dx}{ds}\,\zeta\right) = -\left(\frac{d\zeta}{ds}\middle|\widehat{\alpha\,\zeta}\right) = \left(\nu\middle|\frac{d\zeta}{ds}\right),$$

and, finally, according to Chapter V, (5),

$$\frac{1}{r} = -\frac{(dx|d\zeta)}{ds^2} = -\left(\frac{dx}{ds}\middle|\frac{d\zeta}{ds}\right) = -\left(\alpha\middle|\frac{d\zeta}{ds}\right).$$

Thus, we obtain, as the final forms of the desired equations,

$$\begin{aligned}
\frac{d\alpha}{ds} &= \qquad\quad \frac{\nu}{\rho} + \frac{\zeta}{r}, \\
\frac{d\nu}{ds} &= -\frac{\alpha}{\rho} \qquad\ - \frac{\zeta}{\tau}, \qquad (28) \\
\frac{d\zeta}{ds} &= -\frac{\alpha}{r} + \frac{\nu}{\tau} \qquad\quad .
\end{aligned}$$

It is to be noted that if the curve C is a line of curvature, an asymptotic line, or a geodesic, one of the three quantities $1/\rho$, $1/\tau$, $1/r$ is zero, and the equations become even more simple.

Suppose, now, that C is a curve along which two surfaces are tangent and assume that the normals to the surfaces in the points of C are similarly directed. Then, α, ν, ζ and s, and hence $d\alpha/ds$, $d\nu/ds$, $d\zeta/ds$ are the same whether we think of C as lying on the one surface or on the other. It follows, therefore, from equations (28), that $1/\rho$, $1/\tau$, $1/r$ are the same for both surfaces.

THEOREM 1. *If two surfaces are tangent to one another along a curve and the normals to them in the points of the curve are similarly directed, the curve has the same geodesic curvature and the same geodesic torsion with respect to both surfaces and the surfaces have the same normal curvature in the direction of the curve.*

EXERCISES

1. Show that $\beta/R = \nu/\rho + \zeta/r$.

2. Find the squares of the lengths of the vectors $d\alpha/ds$, $d\nu/ds$, and $d\zeta/ds$.

3. The normals to a surface in the points of a (real) curve are all parallel if and only if the curve is an asymptotic line all of whose points are parabolic or planar. See § 44, Ex. 3.

4. The vector ν for a curve, not a straight line, has always the same direction when and only when the curve is a plane geodesic. See § 52, Ex. 4.

5. Prove that the sum of the squares of the geodesic curvature and geodesic torsion of a curve on the unit sphere is constant if and only if the curve is the tangent indicatrix of a helix.

6. What does the theorem in the text become when the normals to the two surfaces in the points of the curve are oppositely directed?

7. Derive the Frenet-Serret formulas from equations (28).

CHAPTER VIII

MAPPING OF SURFACES

58. Conformal, area-preserving, and isometric maps. Two surfaces, S and $\bar{S}$, are said to be mapped upon one another if there exists a one-to-one correspondence between the points of the one surface and the points of the other. If S and $\bar{S}$ have the parametric representations $x = x(u, v)$ and $\bar{x} = \bar{x}(\bar{u}, \bar{v})$, the most general map of S on $\bar{S}$ is defined by equations, $\bar{u} = \bar{u}(u, v)$, $\bar{v} = \bar{v}(u, v)$, which establish a one-to-one correspondence between the permissible pairs of values of u, v and the permissible pairs of values of $\bar{u}, \bar{v}$, that is, between the points of S and the points of $\bar{S}$. We assume, in particular, that the functions $\bar{u}(u, v)$, $\bar{v}(u, v)$ are analytic functions of u and v.

The parameters u, v on the surface S may be introduced as parameters on the surface $\bar{S}$ by setting $\bar{u} = \bar{u}(u, v)$ and $\bar{v} = \bar{v}(u, v)$ in $\bar{x} = \bar{x}(\bar{u}, \bar{v})$. Then, corresponding points of the two surfaces have the same curvilinear coordinates, and the parametric equations of the surfaces,

$$(1) \qquad x = x(u, v), \qquad \bar{x} = \bar{x}(u, v),$$

serve, not only to define the surfaces, but also the map which has been established between them. Accordingly, we may speak of these equations as the equations of the map of the one surface upon the other.

Conformal maps. A map is said to be conformal if it preserves angles. More explicitly, the map of S on $\bar{S}$ is conformal if the angle between two directed curves through a point P of S is equal always to the angle between the two corresponding directed curves through the corresponding point $\bar{P}$ of $\bar{S}$.

THEOREM 1. *A necessary and sufficient condition that the map (1) be conformal is that the linear elements of S and $\bar{S}$ be proportional:*

$$(2) \qquad \bar{E}du^2 + 2\,\bar{F}dudv + \bar{G}dv^2 = \rho^2(Edu^2 + 2\,Fdudv + Gdv^2),$$

or, what is the same thing, that $\bar{E} : \bar{F} : \bar{G} = E : F : G.$

Suppose that the map is conformal, and consider two arbitrarily chosen corresponding directed curves, C and $\bar{C}$, passing through the arbitrarily chosen corresponding points, P and $\bar{P}$. Since corresponding points of S and $\bar{S}$ have the same curvilinear coordinates (u, v), C and $\bar{C}$ are represented by the same equation in u and v, and hence have at P and $\bar{P}$, respectively, the same direction coordinate dv/du. On the other hand, the elements of arc, ds and $d\bar{s}$, of C and $\bar{C}$ are, in general, different.

If θ and $\bar{\theta}$ are the angles which two other corresponding directed curves through P and $\bar{P}$, with elements of arc δs and $\delta\bar{s}$ and common direction coordinate $\delta v/\delta u$, make respectively with C and $\bar{C}$, then, by Chapter IV, (14),

$$\sin\theta = \frac{D|du\delta v - dv\delta u|}{ds\delta s}, \quad \sin\bar{\theta} = \frac{\bar{D}|du\delta v - dv\delta u|}{d\bar{s}\delta\bar{s}}.$$

Since $\theta = \bar{\theta}$,

$$\frac{D}{ds\delta s} = \frac{\bar{D}}{d\bar{s}\delta\bar{s}}. \tag{3}$$

If the directed u-curves through P and $\bar{P}$, which surely correspond, are taken as the second pair of curves, $\delta s = E^{1/2}du$ and $\delta\bar{s} = \bar{E}^{1/2}du$, and (3) becomes

$$\frac{d\bar{s}}{ds} = \frac{\bar{D}E^{1/2}}{D\bar{E}^{1/2}} = \rho(u, v).$$

Hence, for the elements of arc, ds and $d\bar{s}$, of C and $\bar{C}$ at P and $\bar{P}$, we have

$$d\bar{s} = \rho ds. \tag{4}$$

But P and $\bar{P}$ were any two corresponding points of S and $\bar{S}$, and C and $\bar{C}$ were any two corresponding curves through P and $\bar{P}$. Hence, equation (4) is identically satisfied.

Conversely, if $\rho(u, v)$ exists so that (4) is an identity, the map is conformal. By hypothesis, $d\bar{s} = \rho ds$, and $\delta\bar{s} = \rho\delta s$. If, then, we can show that $\bar{D} = \rho^2 D$, it will follow from (3) that $\sin\theta = \sin\bar{\theta}$, that is, that angle is always preserved. But, we have, from (4), $\bar{E} = \rho^2 E$, $\bar{F} = \rho^2 F$, $\bar{G} = \rho^2 G$, and hence $\bar{D} = \rho^2 D$. Thus, the theorem is proved.

Identity (4) tells us that it is characteristic of a conformal map that corresponding infinitesimal distances at corresponding points are proportional, the factor of proportionality, ρ, depending only on the pair of corresponding points chosen.

THEOREM 2. *A conformal map preserves isometric systems of curves and isometric parameters. Conversely, a map which orders to at least one isometric system of curves and associated isometric parameters a similar system and similar parameters, is conformal.*

If the parameters u, v on S are isometric, the linear element of S is of the form $ds^2 = \lambda(du^2 + dv^2)$. Hence, the map (1) of S on $\bar{S}$ is conformal, by Theorem 1, when and only when the linear element of $\bar{S}$ is of the form $d\bar{s}^2 = \bar{\lambda}(du^2 + dv^2)$. But this form characterizes u, v as isometric parameters on $\bar{S}$, and the theorem is proved.

It is always possible to establish a map which orders to a given isometric system, together with isometric parameters, a prescribed isometric system, together with isometric parameters. We have merely to introduce the isometric system on each surface as the system of parametric curves, employing isometric parameters in each case, and denoting them in both cases by u, v. Then, the map in which corresponding points have the same coordinates (u, v) has the desired property. For, the parametric curves, which surely correspond, constitute now the given and prescribed isometric systems, and u, v are isometric parameters common to these systems.

The map which we have thus established is, by Theorem 2, conformal. Consequently, since there are infinitely many isometric systems of curves on a surface, we conclude that *two surfaces can be mapped conformally upon one another in infinitely many ways.*

In the map of an ordinary surface, $x = x(u, v)$, on its spherical representation, $\zeta = \zeta(u, v)$, corresponding points have the same coordinates (u, v). Hence, the map is conformal when and only when the linear element, $d\sigma^2$, of the sphere is proportional to the linear element, ds^2, of the surface. According to § 50,

$$K ds^2 + K'(dx \mid d\zeta) + d\sigma^2 = 0.$$

Hence, $d\sigma^2 = k^2 ds^2$ if and only if

$$(K + k^2) ds^2 + K'(dx \mid d\zeta) = 0.$$

Unless the second fundamental form, $-(dx \mid d\zeta)$, of the surface is proportional to the first, when the surface is a sphere (§ 37), this relation can hold when and only when $K' = 0$ and $k^2 = -K$. Hence, *the only surfaces, other than spheres, which are mapped conformally upon their spherical representations are the minimal surfaces.*

Area-preserving maps. A map which preserves areas is called area-preserving, or equivalent, or equiareal.

THEOREM 3. *The map* (1) *is area-preserving if and only if* $\bar{D} = D$.

For, the area of a region on S is equal to the area of the corresponding region on $\bar{S}$ if and only if (§ 33)

$$\int\int D du dv = \int\int \bar{D} du dv,$$

where both integrals are extended over the pairs of values of (u, v) which yield the points of the two regions. But, a necessary and sufficient condition that the equality hold for every pair of corresponding regions is, evidently, that $D = \bar{D}$.

Area-preserving maps, though of importance in certain connections, have not the general interest which pertains to conformal maps or the maps yet to be discussed. Suffice it to say, then, that any two surfaces can be mapped upon one another so that areas are preserved.

Isometric maps. These are the maps which preserve distance. Clearly, if every length of arc on S is equal to the corresponding length of arc on $\bar{S}$, the linear element of S is equal to the linear element of $\bar{S}$, and conversely.

THEOREM 4. *The map* (1) *is isometric if and only if the surfaces S and $\bar{S}$ have the same linear element:*

$$\bar{E} du^2 + 2\bar{F} du dv + \bar{G} dv^2 = E du^2 + 2 F du dv + G dv^2,$$

or, in other words, if and only if $\bar{E} = E$, $\bar{F} = F$, $\bar{G} = G$.

It follows that an isometric map is both conformal and area-preserving. The converse is true. For, if the map (1) is conformal, $d\bar{s}^2 = \rho^2 ds^2$ and $\bar{D} = \rho^2 D$; if it is also area-preserving, $\bar{D} = D$, and hence $\rho^2 = 1$ and $d\bar{s}^2 = ds^2$. Thus:

THEOREM 5. *A necessary and sufficient condition that a map be isometric is that it be conformal and area-preserving.*

In other words, a map which preserves lengths also preserves angles and areas; and, a map which preserves both angles and areas also preserves lengths.

As we shall see shortly, it is not always possible to map one of two given surfaces isometrically upon the other.

EXERCISES

1. If a map is not conformal, there exists a unique orthogonal system of curves on each surface to which corresponds an orthogonal system on the other surface.

2. If a map of two ordinary surfaces upon one another preserves asymptotic lines, it also preserves conjugate systems.

3. If a map of two ordinary surfaces upon one another does not preserve a family of asymptotic lines, there exists a unique conjugate system on each surface to which corresponds a conjugate system on the other surface.

4. If the asymptotic lines on one of two ordinary surfaces which are mapped on one another correspond to a conjugate system on the other, the asymptotic lines on the second surface correspond to a conjugate system on the first.

5. Prove that all the surfaces which are parallel to the surface $x = x(u, v)$, that is, have the same normals, are given by $y = x + a\zeta$, where a is an arbitrary constant.

6. Determine in the map of an ordinary surface $S: x = x(u, v)$ on a parallel surface $\bar{S}: \bar{x} = x + a\zeta$ the orthogonal (conjugate) system on each surface which corresponds to an orthogonal (conjugate) system on the other. Show that a necessary and sufficient condition that there exists a surface $\bar{S}$ such that the asymptotic lines on S correspond to a conjugate system on $\bar{S}$ is that S be of constant mean curvature, not zero.

7. The surface $x = x(u, v)$ is carried by the inversion in the sphere with the origin as center and radius a into the surface $\bar{x} = a^2x/(x|x)$. Show that the map thus established between the two surfaces is conformal and preserves lines of curvature.

8. Show that the affine transformation of the plane,

$$\bar{x}_1 = a_1x_1 + a_2x_2 + a_3, \quad \bar{x}_2 = b_1x_1 + b_2x_2 + b_3, \quad a_1b_2 - a_2b_1 \neq 0,$$

always preserves the ratios of areas. When does it preserve areas themselves?

9. Show that a catenoid can be mapped isometrically on a right helicoid.

10. If the normals at corresponding points of the surfaces (1) are parallel, there is, in general, a unique system of curves on each surface such that the curves of the two systems correspond and have at corresponding points parallel tangents, and these systems are the corresponding conjugate systems of the map.

Suggestion. First show that functions a, b, c, d of u, v exist such that $y_u = ax_u + bx_v$, $y_v = cx_u + dx_v$, $ad - bc \neq 0$. Then base the discussion on these equations, simplifying them, after the first fact has been established, by a convenient choice of parametric curves.

59. The absolute properties of a surface. Applicability. We have just seen that the map (1) is isometric if and only if the two surfaces, S and $\bar{S}$, have the same first fundamental form, $Edu^2 + 2\,Fdudv + Gdv^2$. Hence, an isometric map preserves every geometric magnitude which is expressible in terms of E, F, G and their partial derivatives with respect to u and v. Distance, angle, and area are evidently magnitudes of this type. Total curvature and geodesic curvature are also, for we have shown that both are expressible in the manner described (§§ 49, 51).

On the other hand, the mean curvature of a surface, the normal curvature and the geodesic torsion at a point in a given direction, and the ordinary curvature of a curve on the surface are geometric magnitudes which depend, not only on E, F, G, but also on e, f, g, and hence are not, in general, preserved by an isometric map.

It is evident, for the same reason, that an isometric map does not, in general, carry a line of curvature into a line of curvature, or an asymptotic line into an asymptotic line, or a conjugate system of curves into a conjugate system of curves. It does, however, carry an isometric system into an isometric system, and a geodesic into a geodesic.

With a view to finding a geometric distinction between the properties preserved by isometric maps and those which are, in general, changed, let us tabulate them.

A. Properties preserved: distance, angle, area, total curvature, geodesic curvature, geodesics, isometric systems.

B. Properties changed: mean curvature, normal curvature, geodesic torsion, curvature and torsion of a curve on the surface, lines of curvature, asymptotic lines, conjugate systems.

Into all the properties B there enters, directly or indirectly, either the concept of the curvature, or that of the torsion, of a space curve. Now, the curvature and torsion of a space curve cannot be measured without getting off the curve. In other words, they are not properties of the curve in itself, but properties of the curve in its relationship to the space in which it is imbedded. Hence, all the properties B pertain, not to the surface in itself, but to the surface in its relationship to surrounding space.

On the other hand, properties A are all possible of definition on the surface, and are actually independent of the relationship of the surface to the space in which the surface is imbedded. This is intuitively evident in the case of distance, angle, and area; it is unnecessary to leave the surface to measure any of these magnitudes. Similarly, it is true of isometric systems and geodesics, inasmuch as both may be defined in terms of distance. Total curvature and geodesic curvature, however, were defined with reference to the space surrounding the surface and it might appear that they belong more properly with properties B. But both, like the rest of properties A, depend only on the first fundamental form of the surface, and hence there is good reason to believe that it is really unnecessary to leave the surface in order to define them. As a matter of fact, this is the case, as we shall show later.

Properties A and all other properties which pertain merely to the surface, regardless of the surrounding space, we shall call the *absolute properties* of the surface. They make up the *geometry on the surface* in the sense of Gauss, who was the first to investigate them systematically.

Two surfaces may have the same absolute geometry and yet look entirely different to an observer who views them from surrounding space. A plane and a cylinder certainly do not look alike, and yet they have the same absolute geometry.

In this connection, it is essential that the reader recall that we are dealing only with properties of a surface in the small (§ 1). Accordingly, we mean by the foregoing statement, not that the plane and the cylinder in their unrestricted extents have the same absolute geometry, but that sufficiently restricted portions of them have the same absolute properties.

Applicability. We seek now the geometric operations which, when applied to the surface, change the properties of the surface relative to the surrounding space, but preserve the absolute properties. Certainly, the rigid motions of space are not to be included among these operations, since they preserve both types of properties. Consequently, the operations must actually deform the surface, in some such way as the plane of the foregoing example may be deformed into the cylinder.

The absolute properties of a surface depend, as we have seen, only on distance. Accordingly, the deformations of a surface in which we are interested are simply those which preserve distance.

A deformation which stretches, compresses, or tears the surface obviously changes distance. There is left, then, only the deformations whose sole effect is the bending of the surface. Thus, the deformations of a surface which preserve its absolute geometry are those which bend but do not stretch, compress, or tear the surface.

For example, a cylinder, or better, a portion thereof can be deformed by bending, without stretching, compression, or tearing, into a plane.

DEFINITION. *Two surfaces, each of which can be deformed into the other by bending, without stretching, compression, or tearing, are said to be applicable to one another.*

The process of deforming one of two applicable surfaces into the other establishes a map of the one surface on the other. Since distance is preserved, this map is isometric. Hence, the applicability of two surfaces implies the existence of an isometric map of one on the other.

The converse is, to all intents and purposes, true. E. E. Levi has shown that, if two surfaces are mapped isometrically, then one is applicable either to the other or to a surface symmetric to the other, that is, to a surface which is the reflection of the other in a point or a plane.*

Henceforth, we may, then, think of applicability and the existence of an isometric map as equivalent.

The fact that an isometric map preserves total curvature may be stated as follows:

THEOREM 1. *A necessary condition that two surfaces be applicable is that it be possible to establish between their points a one-to-one correspondence so that the total curvatures at corresponding points are equal.*

The condition is not, in general, sufficient. It is sufficient, however, in case the total curvatures are constant.

THEOREM 2. *Two surfaces of the same constant total curvature are always applicable.*

For the present, we content ourselves with a proof of the theorem in the case in which the total curvature is zero. We have, then, to show that each two developable surfaces, including planes, are applicable to one another.

* See Bianchi, *Vorlesungen über Differentialgeometrie*, second edition, p. 178.

It suffices to prove that every developable is applicable to a plane. The fact is obvious in the case of a cylinder or a cone. In discussing the case of the tangent surface of a twisted curve C_1, we think of C_1 as represented by the intrinsic equations

$$C_1: \qquad \frac{1}{R} = f(s), \qquad \frac{1}{T} = \phi(s),$$

and consider in conjunction with it the plane curve,

$$C_0: \qquad \frac{1}{R} = f(s), \qquad \frac{1}{T} = 0.$$

The parametric representations $y^{(1)} = x^{(1)} + r\alpha^{(1)}$ and $y^{(0)} = x^{(0)} + r\alpha^{(0)}$ of the tangent surfaces, S_1 and S_0, of C_1 and C_0, since they are in terms of the same parameters r, s, establish a map of S_1 on S_0. Computation shows that S_1 and S_0 have the same linear element, namely,

$$\left(1 + \frac{r^2}{R^2}\right) ds^2 + 2\, drds + dr^2. \tag{5}$$

Hence, the map is isometric and S_1 is applicable to S_0. But S_1 is the given developable and S_0 is a plane.

We can go further and establish Levi's theorem in this case. Consider the variable curve

$$C_\eta: \qquad \frac{1}{R} = f(s), \qquad \frac{1}{T} = \eta\phi(s),$$

where η is a parameter. The tangent surface, S_η, of C_η has the linear element (5), and hence is always isometrically mapped on S_1 or S_0. But, when η decreases continuously from 1 to 0, S_η varies continuously from S_1 to S_0. Thus, we have actually before us a continuous bending, without stretching or tearing, of S_1 into S_0.

We have already noted that an isometric map preserves the property that a curve be a geodesic. In other words:

THEOREM 3. *A necessary condition that two surfaces be applicable is that they can be mapped geodesically, that is, so that to a geodesic on the one surface corresponds always a geodesic on the other surface.*

This condition, too, is not, in general, sufficient, as is evident from the following result, due to Dini.

THEOREM 4. *If two surfaces are mapped geodesically, then either (a) the map is homothetic, or (b) the linear elements of the surfaces can be reduced simultaneously to the forms*

$$(6)\quad ds^2 = (U - V)(du^2 + dv^2), \quad d\bar{s}^2 = \left(\frac{1}{V} - \frac{1}{U}\right)\left(\frac{du^2}{U} + \frac{dv^2}{V}\right),$$

*where $U = U(u)$ and $V = V(v)$.**

A map is said to be homothetic if corresponding distances are always in the same ratio, that is, if the linear element of the one surface is a constant multiple of the linear element of the other: $d\bar{s}^2 = c^2ds^2$. It is readily seen, then, by application of the Gauss equation, that $\bar{K} = K/c^2$, where K and $\bar{K}$ are the total curvatures of the two surfaces at corresponding points.

Though neither of the necessary conditions of Theorems 1 and 3 is, in itself, sufficient, the two conditions together are sufficient.

THEOREM 5. *A necessary and sufficient condition that two surfaces be applicable is that they can be mapped geodesically so that the total curvatures in corresponding points are equal.*

We give, in outline, the proof of the sufficiency of the condition. Since the map is geodesic, it is in any case of one of the two types described in Dini's theorem. If it is of type (a), the added prescription that $\bar{K} = K$ implies that $d\bar{s}^2 = ds^2$ and the surfaces are applicable. If it is of type (b), it can be shown, though not without difficulty, that the requirement that $\bar{K} = K$ makes the common total curvature of the surfaces constant. Hence, by Theorem 2, the two surfaces are actually applicable to one another.

A surface whose linear element is reducible to either of the forms (6) is known as a surface of Liouville. We have just seen that these surfaces comprise, among others, the surfaces of constant curvature. Hence, Dini's theorem would lead us to believe that *any two surfaces of constant curvature can be mapped geodesically*. To establish this important proposition, it would suffice to show that any surface of constant curvature can be mapped geodesically on a plane. As a matter of fact, Beltrami

* A proof of the theorem will be found in Darboux, *Leçons sur la théorie générale des surfaces*, Vol. 3, p. 49.

has proved, not only that the surfaces of constant total curvature can be mapped geodesically on the plane, but also that they are the only surfaces with this property.*

By means of the theory of applicability, we may establish the theorem of § 28 to the effect that a curve on a developable surface, other than a straight line, is rectified by rolling the surface out on a plane if and only if the developable surface is the rectifying developable of the curve. For, the curve is rectified when and only when it is a geodesic on the surface, and we know, from § 52, Theorem 1, that it is a geodesic if and only if the surface is its rectifying developable.

60. Applicability of surfaces of constant curvature. To show that two surfaces of the same constant total curvature are applicable, we shall prove that the linear element of an arbitrary surface of given constant curvature is reducible to a form which depends merely on the curvature.

Let P be an arbitrarily chosen point on the surface and let C be an arbitrarily chosen geodesic passing through P. Take C as the parametric curve $u = 0$ and let v be the arc of C, measured from P. Choose as the u-curves the geodesics orthogonal to C, taking u as their common arc, measured from C. Then u and v are geodesic parameters and the linear element of the surface has the form

$$ds^2 = du^2 + G dv^2. \tag{7}$$

The element of arc and the geodesic curvature of the curve C: $u = 0$ are $(\sqrt{G})_0 dv$ and $(\partial \log \sqrt{G}/\partial u)_0$, where the subscript zero indicates that the expressions have been evaluated for $u = 0$. Since the arc of C is v and C is a geodesic, it follows that

$$(\sqrt{G})_0 = 1, \qquad \left(\frac{\partial \sqrt{G}}{\partial u}\right)_0 = 0. \tag{8}$$

According to § 53, we have, for the determination of $\sqrt{G}$, the differential equation

$$\frac{\partial^2 \sqrt{G}}{\partial u^2} + K\sqrt{G} = 0, \tag{9}$$

subject, of course, to the initial conditions (8).

* For a proof, see Bianchi, *Vorlesungen über Differentialgeometrie*, second edition, p. 442.

$K > 0$. The general solution of the equation (9) in this case is

$$\sqrt{G} = a(v) \cos(\sqrt{K}\, u) + b(v) \sin(\sqrt{K}\, u).$$

The "constants of combination" a and b are functions of v inasmuch as $\sqrt{G}$ depends theoretically on v as well as on u. Applying the initial conditions (8), we find that $a(v) = 1$ and $b(v) = 0$. Hence, $\sqrt{G} = \cos(\sqrt{K}\, u)$, and the linear element becomes

$$ds^2 = du^2 + \cos^2(\sqrt{K}\, u)dv^2. \tag{10a}$$

$K < 0$. Here, the general solution of (9),

$$\sqrt{G} = a(v) \cosh(\sqrt{-K}\, u) + b(v) \sinh(\sqrt{-K}\, u),$$

when it is subjected to the initial conditions (8), reduces to $\sqrt{G} = \cosh(\sqrt{-K}\, u)$, so that the linear element becomes

$$ds^2 = du^2 + \cosh^2(\sqrt{-K}\, u)dv^2. \tag{10b}$$

$K = 0$. In this case the general solution of (9), namely, $\sqrt{G} = a(v)u + b(v)$, reduces to $\sqrt{G} = 1$ when the initial conditions are applied. Hence, the linear element takes the form

$$ds^2 = du^2 + dv^2. \tag{10c}$$

We remark, next, that the parameters u and v are determined, except for sign, once the point P and the geodesic C have been chosen. Consequently, since there are ∞^2 points on a surface and ∞^1 geodesics through a point, the linear element of the given surface can be put into one of the forms (10) in ∞^3 ways.

Consider, now, any two surfaces, S and $\bar{S}$, of the same constant curvature. Let P be a given point on S and C a given geodesic through P. Then $\bar{S}$ can be applied to S so that an arbitrarily chosen one of the ∞^2 points on S coincides with P and an arbitrarily chosen one of the ∞^1 geodesics through this point coincides with C. Thus, we come to the following conclusion.

THEOREM 1. *Two surfaces of the same constant total curvature are applicable to one another in* ∞^3 *ways.*

The two surfaces have, then, the same absolute geometry. Accordingly, if we know the absolute geometry on one surface of given constant curvature, we know that of every other surface of the same curvature.

As the typical surface of constant positive curvature, $1/a$, it is natural to take the sphere of radius a.

As the typical surface of constant negative curvature, it is customary to take the surface of revolution obtained by revolving the tractrix (Fig. 28) about its asymptote. The tractrix is characterized by the fact that the segment PQ of its tangent is always of the same length. If this constant length is a, the curvature of the surface is $-1/a^2$.

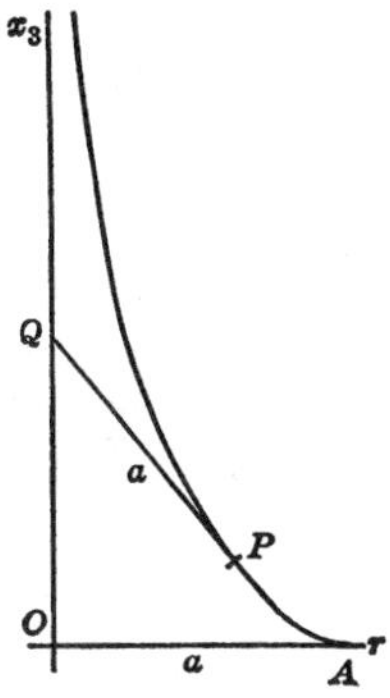

FIG. 28

Since our theory of applicability applies merely to restricted portions of surfaces, the foregoing remarks apply to absolute geometry in the small, and not to absolute geometry in the large. In fact, two surfaces of the same constant curvature have, in general, different absolute geometries in the large. For example, through two points of a circular cylinder there pass infinitely many geodesics, whereas through two points of a plane there passes just one geodesic.

It follows from Theorem 1 that a surface of constant total curvature is applicable to itself in ∞^3 ways. In the case of a plane, it is evident that no bending is necessary; the ∞^3 operations of applicability may be taken as the ∞^3 rigid motions of the plane into itself. Similarly, the ∞^3 rotations of a sphere about its center effect the applicability of the sphere to itself.

61. Continuous deformations of surfaces of variable curvature. We seek the surfaces of variable curvature, K, which are applicable to themselves in a continuous infinity of ways, or, as we say, admit continuous deformations into themselves. A surface of revolution is an example of a surface of this type, for it is carried into itself by a continuous rotation about its axis, that is, by the ∞^1 rotations about its axis; thereby, the paths traced by the individual points of the surface are the parallels. Again, a right helicoid admits a continuous screw motion into itself in which the path curves are the circular helices on the helicoid.

In these cases and, in fact, in every case, *the path curves of the continuous deformation*, since the deformation preserves the total curvature, *are the curves $K =$ const. on the given surface.*

Consider, now, an arbitrary, but fixed, one of the isometric transformations which constitute the continuous deformation of the surface into itself. *This transformation, T, carries each point on a chosen curve $K =$ const. a fixed distance along this curve.* For, T is contained in the continuous deformation, carries each curve $K =$ const. into itself, and preserves distance. Hence, it carries every directed distance measured along a chosen curve $K =$ const. into an equal directed distance along this curve. But this is possible only if it transports each point of the curve along the curve through a fixed distance.

It is clear from the foregoing examples that the fixed distance varies with the curve $K =$ const. chosen. However, if it is known for just one curve, it is determined for all. In other words:

THEOREM 1. *A constituent isometric transformation, T, of a continuous deformation of a surface into itself is uniquely determined if the distance, d, through which it transports the points of just one of the curves $K =$ const. is known.*

By hypothesis, T carries an arbitrarily chosen point P_0 on the preferred curve $K =$ const. into a specific point P_0' on this curve. Hence, since it is conformal, it must carry the orthogonal trajectory, C, of the curves $K =$ const. which passes through P_0 into the orthogonal trajectory, C', of the curves $K =$ const. which passes through P_0' (Fig. 29). Consequently, it must carry an arbitrarily chosen point P on C into the point P' on C' in which C' is met by the curve $K =$ const. which passes through P. Thus, the effect of T on every point on the surface is fixed, and T is uniquely determined.

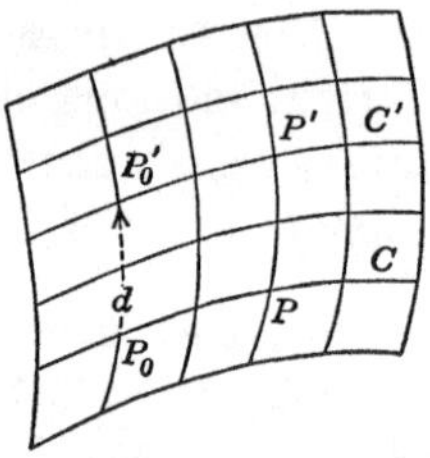

FIG. 29

We now introduce the curves $K =$ const. as the v-curves and their orthogonal trajectories as the u-curves, and, in particular, take, as v, the arc of the preferred curve $K =$ const. By the hypothesis of Theorem 1, if (u_0, v) are the coordinates of P_0, $(u_0, v + d)$ are those of P_0'. Hence, if P has the coordinates (u, v), those of P' are $(u, v + d)$. In other words, the transformation T, thought of as carrying $P: (u, v)$ into $P': (u', v')$, has the equations $u' = u$, $v' = v + d$.

Since T was an arbitrary one of the transformations which constitute the given continuous deformation, the equations $u' = u$, $v' = v + d$, when d takes on all values in a certain interval including $d = 0$, represent the deformation. Hence:

THEOREM 2. *A continuous deformation of a surface of variable curvature depends on one parameter. By a proper choice of curvilinear coordinates, its equations are reducible to the form* $u' = u$, $v' = v + d$, *where* d *is the parameter.*

When the points P_0' and P', starting from the positions P_0 and P, are transported by the deformation along the v-curves ($K =$ const.) through P_0 and P, the distance $P_0'P'$ measured along C' remains always the same as the distance P_0P along C. It follows, then, that each two v-curves cut equal segments from their orthogonal trajectories,—the u-curves. Hence, the u-curves are geodesics and the v-curves are geodesic parallels.

We may now choose, as u, the common arc of the u-curves. Then u, v are geodesic parameters and the linear element of the surface has the form $du^2 + Gdv^2$.

Since the surface admits the continuous deformation $u' = u$, $v' = v + d$, the linear element formed for the point P': $(u, v + d)$, namely $du^2 + G(u, v + d)dv^2$, must be identical with the linear element, $du^2 + G(u, v)dv^2$, formed for the point P: (u, v); that is, $G(u, v + d)$ must be equal to $G(u, v)$ for all values of d. But this is possible if and only if G is a function of u alone. Thus, we have arrived at the following conclusion.

THEOREM 3. *A necessary and sufficient condition that a surface admit a continuous deformation into itself is that its linear element be reducible to the form*

$$ds^2 = du^2 + U(u)dv^2. \tag{11}$$

Comparison of (10a), (10b), and (10c) with (11) shows that the theorem holds for a surface of constant curvature as well as for a surface of variable curvature.

If the linear element of a surface is in the form (11), the family of v-curves, besides consisting of geodesic parallels, is an *isometric family*, that is, forms with its orthogonal trajectories an isometric system. Conversely, if there exists on a surface an isometric family of geodesic parallels, the linear element is reducible to the form (11), as may readily be shown. Thus, we may restate Theorem 3 as follows:

THEOREM 4. *A surface admits a continuous deformation into itself if and only if it contains an isometric family of geodesic parallels.*

In case the curvature of the surface is variable, the curves of the family are the path curves of the deformation and hence are the curves $K = \text{const}$.

THEOREM 5. *Two mutually applicable surfaces of variable curvature are applicable in a continuous infinity of ways if and only if one, and hence each, of them admits a continuous deformation into itself.*

For, there exists a continuous infinity of isometric maps of the one surface upon the other if and only if, after the one has been deformed into the other, the single resulting surface admits a continuous deformation into itself.

The continuous infinity of isometric maps depends on one parameter, namely the parameter of the continuous deformation. Hence, two surfaces of variable curvature are mutually applicable in at most ∞^1 ways.

According to Theorem 5, every surface which is applicable to a surface of revolution admits a continuous deformation into itself. Moreover, the converse is true:

THEOREM 6. *Every surface which admits a continuous deformation into itself is applicable to a surface of revolution.*

We shall prove the theorem in the general case later (§69). For the moment we content ourselves with a proof in the special case of the right helicoid. Based on the usual parametric equations, $x_1 = u \cos v$, $x_2 = u \sin v$, $x_3 = av$, the linear element of the helicoid is

$$ds^2 = du^2 + (u^2 + a^2)dv^2. \tag{12}$$

Incidentally, we note that this is of the form (11).

Parametric equations of a surface of revolution whose axis is the x_3-axis are $x_1 = r \cos v$, $x_2 = r \sin v$, $x_3 = f(r)$, where r, v are polar coordinates in the (x_1, x_2)-plane and $x_3 = f(r)$ represents the plane curve whose rotation generates the surface (Fig. 30). The r-curves are the meridians and the v-curves, the parallels, on the surface.

The linear element of the surface is found to be

$$ds^2 = (1 + f'^2)\, dr^2 + r^2 dv^2,$$

where $f' = df/dr$. When we set

$$du = \sqrt{1 + f'^2}\, dr, \tag{13}$$

it reduces to

$$ds^2 = du^2 + r^2 dv^2. \tag{14}$$

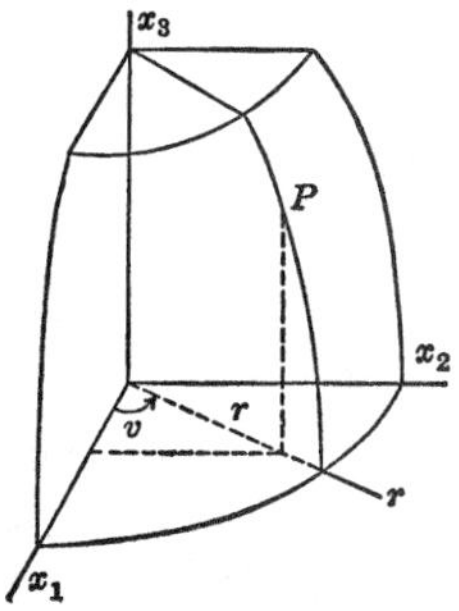

FIG. 30

Here, r is of course a function of u, the inverse of the function $u = \phi(r)$ defined by (13).

Comparison of (14) with (12) shows that, if it is possible to determine $f(r)$ so that $r^2 = u^2 + a^2$ or $u^2 = r^2 - a^2$, we will have at hand a surface of revolution to which the helicoid is applicable. Setting $u^2 = r^2 - a^2$ in (13), we readily find that a suitable choice of $f(r)$ is $a \cosh^{-1}(r/a)$. But the equation $x_3 = a \cosh^{-1}(r/a)$ or $r = a \cosh(x_3/a)$ represents, in the (x_3, r)-plane, a catenary with the axis of x_3 as axis.

Thus, a right helicoid may be deformed into the surface of revolution obtained by revolving a catenary about its axis, a so-called catenoid. Thereby, the helices on the helicoid go over into the parallels on the catenoid.

EXERCISES

1. Prove that the surface

$$x_1 = a(\cos u + \cos v), \qquad x_2 = a(\sin u + \sin v), \qquad x_3 = c(u + v), \qquad ac \neq 0$$

admits a continuous deformation into itself. Begin by showing that the curves K = const. are the curves $u - v$ = const., and then introduce these curves and their orthogonal trajectories as the parametric curves. Show finally that the surface is a right helicoid.

2. Show that an isometric family of geodesic parallels on a sphere consists necessarily of circles lying in planes perpendicular to a diameter of the sphere. See § 49, Ex. 2.

CHAPTER IX

THE ABSOLUTE GEOMETRY OF A SURFACE

62. Geodesic polar coordinates. For the purposes of this chapter, it will be necessary to develop a new type of curvilinear coordinates on the surface, similar to polar coordinates in the plane. As the basis for these coordinates, we take the geodesics which pass through a given regular point O of the surface. It can be shown that there exists a region about O such that there is just one of these geodesics which goes through an arbitrarily chosen point of the region, other than O, before passing out of the region.* Hence, in this region, to which we shall restrict ourselves, the geodesics form a family of curves in the sense in which we defined the term.

Consider, in conjunction with the geodesics through O, the geodesic circles with O as center, that is, the loci of points moving at fixed distances, measured along the geodesics, from O. It is reasonable to expect that these geodesic circles cut the geodesics orthogonally, and that the two families of curves can be employed as the basis for a system of polar coordinates on the surface.

In discussing these questions, we shall mean by the geodesics through O merely the half-geodesics issuing from O. When we introduce these geodesics as the u-curves, taking their common arc, measured from O, as the parameter u ($u \geqq 0$), the v-curves ($u =$ const.) are the geodesic circles with O as center.

Since u is the arc of every u-curve, it follows, in the usual way, that $E = 1$. Then the condition, $C^2_{11} = 0$, that the u-curves be geodesics reduces to $F_u = 0$, and $F = V(v)$.

Since $u = 0$ fixes the point O, regardless of the value of v, we have $x(0, v) = c$, where (c_1, c_2, c_3) are the space coordinates of O. Hence $x_v(0, v) = 0$. Now $F = (x_u | x_v)$, and therefore $F = 0$ when $u = 0$. It follows, since F is independent of u, that $F \equiv 0$. Hence, the geodesic circles are the orthogonal trajectories of the geodesics, and u, v are geodesic parameters.

* See Darboux, *Leçons sur la théorie générale des surfaces*, Vol. 2, pp. 408, 409.

The parameter u is definitely fixed as the common arc of the geodesics. It is evident geometrically that we may choose as the parameter v the directed angle at O from a fixed geodesic to an arbitrary geodesic. The curvilinear coordinates u, v on the surface are, then, known as geodesic polar coordinates.

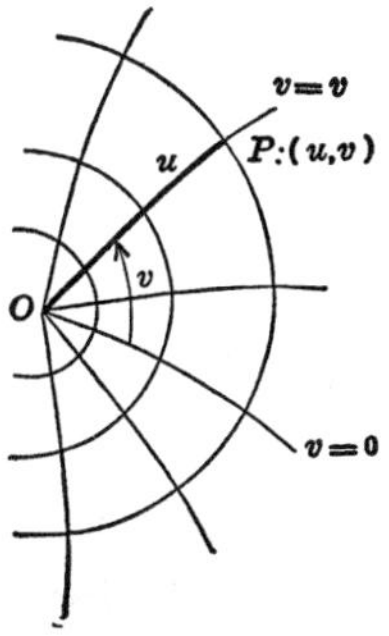

FIG. 31

THEOREM 1. *A necessary and sufficient condition that geodesic parameters u, v be geodesic polar coordinates is that*

$$\lim_{u \to 0} \frac{G}{u^2} = 1. \tag{1}$$

We suppose first that u, v are geodesic polar coordinates and deduce relation (1). Developing the functions $x(u, v)$ in power series in u, we have

$$x(u, v) = x(0, v) + x_u(0, v)u + x_{uu}(0, v)\frac{u^2}{2!} + \cdots,$$

or, since $x(0, v) = c$,

$$x(u, v) = c + \alpha(v)u + \beta(v)\frac{u^2}{2!} + \cdots. \tag{2}$$

Since u is the arc of the geodesic $v = v$, $\alpha(v) = x_u(0, v)$ is the unit vector tangent at O to this geodesic. Hence, if $\Delta\theta$ is the angle at O from the geodesic $v = v$ to the geodesic $v = v + \Delta v$, measured in the direction of increasing v,

$$\sin \Delta\theta = (\alpha \ \alpha + \Delta\alpha | \alpha \ \alpha + \Delta\alpha)^{1/2},$$

where $\Delta\alpha$ is the increment in α due to the increment Δv in v. Then

$$\frac{\sin \Delta\theta}{\Delta\theta}\frac{\Delta\theta}{\Delta v} = \left(\alpha \ \frac{\Delta\alpha}{\Delta v}\middle| \alpha \ \frac{\Delta\alpha}{\Delta v}\right)^{1/2},$$

and

$$\frac{d\theta}{dv} = (\alpha \ \alpha' | \alpha \ \alpha')^{1/2} = [(\alpha|\alpha)(\alpha'|\alpha') - (\alpha|\alpha')^2]^{1/2}.$$

Since $(\alpha|\alpha) = 1$, $(\alpha|\alpha') = 0$ and therefore

$$\frac{d\theta}{dv} = \sqrt{\alpha'|\alpha'}. \tag{3}$$

From (2),

$$x_v(u, v) = \alpha'(v)u + \beta'(v)\frac{u^2}{2!} + \cdots .$$

Thus,

$$G = (\alpha'|\alpha')u^2 + (\alpha'|\beta')u^3 + \cdots ,$$

and

(4) $$\lim_{u \to 0} \frac{G}{u^2} = (\alpha'|\alpha').$$

By hypothesis, $d\theta/dv = 1$. Hence $(\alpha'|\alpha') = 1$, and relation (1) is established.

Conversely, if u, v are geodesic parameters for which (1) holds, they are geodesic polar coordinates. For, in the first place, (1) implies that $G(0, v) = 0$, inasmuch as, if $G(0, v)$ were not zero, G/u^2 would become infinite when u approaches zero. It follows, then, since $G = (x_v|x_v)$, that $x_v(0, v) = 0$, and $x(0, v) = c$. Thus, the geodesics $v =$ const. all go through a point O, and their common arc is measured from this point.

Equation (2) and equations (3) and (4), which follow from (2), are now valid. Since (1) holds, equations (3) and (4) tell us that $d\theta/dv = 1$. Consequently, v is actually the angle at O from the geodesic $v = 0$ to the geodesic $v = v$, and the proof is complete.

Equivalent to the relations $G(0, v) = 0$ and $\lim (G/u^2) = 1$ are the relations $\sqrt{G(0, v)} = 0$ and $\lim (\sqrt{G}/u) = 1$. Thus, we may restate Theorem 1 as follows:

THEOREM 2. *The parameters u, v are geodesic polar coordinates if and only if $ds^2 = du^2 + Gdv^2$, and the power series in u for $\sqrt{G}$ is of the form*

(5) $$\sqrt{G} = u + a_2\frac{u^2}{2!} + a_3\frac{u^3}{3!} + \cdots .$$

Comparing this series with

$$\sqrt{G} = (\sqrt{G})_0 + \left(\frac{\partial\sqrt{G}}{\partial u}\right)_0 u + \left(\frac{\partial^2\sqrt{G}}{\partial u^2}\right)_0 \frac{u^2}{2!} + \left(\frac{\partial^3\sqrt{G}}{\partial u^3}\right)_0 \frac{u^3}{3!} + \cdots,$$

where each of the coefficients is evaluated for $u = 0$, we see that

(6) $$0 = (\sqrt{G})_0,\quad 1 = \left(\frac{\partial\sqrt{G}}{\partial u}\right)_0,\quad a_2 = \left(\frac{\partial^2\sqrt{G}}{\partial u^2}\right)_0,\quad a_3 = \left(\frac{\partial^3\sqrt{G}}{\partial u^3}\right)_0.$$

According to Chapter VII, (10),

$$\frac{\partial^2\sqrt{G}}{\partial u^2} = -K\sqrt{G}, \qquad \frac{\partial^3\sqrt{G}}{\partial u^3} = -K\frac{\partial\sqrt{G}}{\partial u} - \sqrt{G}\frac{\partial K}{\partial u}.$$

Hence, $a_2 = 0$, $a_3 = -K_0$, and (5) becomes

$$\sqrt{G} = u - \frac{1}{6}K_0u^3 + \cdots, \tag{7}$$

where K_0 is the value of the total curvature at O.

We may now compute the length of the circumference, and also the area, of the geodesic circle, $u = u$, whose center is at O, and whose radius is u. For the circumference, we have

$$C = \int_0^{2\pi} \sqrt{G}\, dv = 2\pi u - \frac{\pi}{3}K_0u^3 + \cdots, \tag{8}$$

and for the area

$$A = \int_0^u \int_0^{2\pi} \sqrt{G}\, dudv = \pi u^2 - \frac{\pi}{12}K_0u^4 + \cdots. \tag{9}$$

Hence, the circumference and area, when the radius u is sufficiently small, are less than or greater than $2\pi u$ and πu^2 respectively, according as the total curvature at O is positive or negative.

From (8) and (9) we obtain the expressions,

$$K_0 = \frac{3}{\pi}\lim_{u\to 0}\frac{2\pi u - C}{u^3}, \qquad K_0 = \frac{12}{\pi}\lim_{u\to 0}\frac{\pi u^2 - A}{u^4}, \tag{10}$$

for the total curvature at O. Since these expressions involve merely distance and area, they constitute interpretations of the total curvature which are independent of the space in which the surface is imbedded. Hence, they definitely establish total curvature as an absolute property of the surface.

63. A differential equation of the geodesics in terms of geodesic parameters. Let C: $u = u(s)$, $v = v(s)$ be a curve on the surface, other than a u-curve, and let it be directed in the direction in which v increases. Let θ be the directed angle at an arbitrary point P of C from the directed u-curve through P to the directed curve C. Then, if u, v are geodesic parameters, so that

$$ds^2 = du^2 + Gdv^2,$$

we readily find that

$$\cos\theta = \frac{du}{ds}, \qquad \sin\theta = \sqrt{G}\frac{dv}{ds}. \tag{11}$$

The curve C is a geodesic provided the vector d^2x/ds^2 is perpendicular to each of two nonparallel vectors in the tangent plane (§ 54). Since C is not a u-curve, we may take as these vectors x_u and the unit vector α tangent to C. But d^2x/ds^2 is always perpendicular to α. Consequently, C is a geodesic if and only if $(x_u \,|\, d^2x/ds^2) = 0$, or

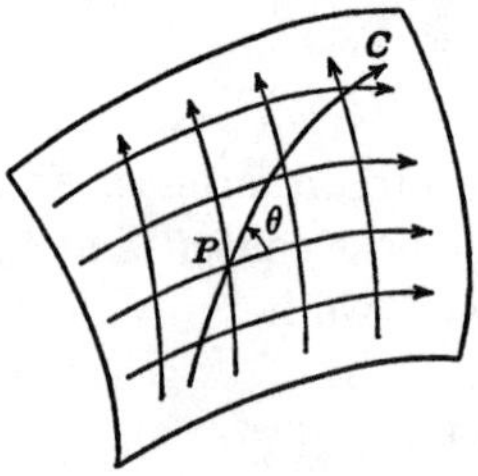

FIG. 32

$$\frac{d^2u}{ds^2} + (x_{uu}\,|\,x_u)\left(\frac{du}{ds}\right)^2 + 2\,(x_{uv}\,|\,x_u)\frac{du}{ds}\frac{dv}{ds} + (x_{vv}\,|\,x_u)\left(\frac{dv}{ds}\right)^2 = 0.$$

Employing relations (4) of Chapter VI, we find that this equation reduces to

$$\frac{d^2u}{ds^2} - \frac{1}{2}G_u\left(\frac{dv}{ds}\right)^2 = 0,$$

and, hence, by virtue of (11), to

$$\frac{d\theta}{ds} + \frac{\partial\sqrt{G}}{\partial u}\frac{dv}{ds} = 0. \tag{12}$$

We assumed that C was not one of the geodesics $v = \text{const.}$ However, (12) holds also for these geodesics, inasmuch as, for them, $dv = 0$ and $d\theta = 0$.

Equation (12) is due to Gauss and will be referred to as the Gauss equation of the geodesics.

64. Geodesic triangles. The sum of the angles of a triangle whose sides are segments of geodesics is closely connected with the integral of the total curvature,

$$T = \iint K\,dA, \tag{13}$$

extended over the triangle. This integral, the *curvatura integra* of Gauss, we shall call the integral curvature of the triangle.

THEOREM 1. *The sum of the angles α_1, α_2, α_3 of a geodesic triangle $A_1A_2A_3$ differs from π by the integral curvature of the triangle:*

$$(14) \qquad \alpha_1 + \alpha_2 + \alpha_3 - \pi = T.$$

In giving the proof, we shall restrict ourselves to the case in which the triangle lies wholly within the region of definition of a system of geodesic polar coordinates whose pole is at one vertex. Let this be the vertex A_1, and choose the polar coordinate system so that the geodesics A_1A_2 and A_1A_3 are respectively $v = 0$ and $v = \alpha_1$.

Since $ds^2 = du^2 + Gdv^2$, we know that

$$K = -\frac{1}{\sqrt{G}}\frac{\partial^2\sqrt{G}}{\partial u^2}, \qquad dA = \sqrt{G}\,dudv.$$

Hence,

$$T = -\int_0^{\alpha_1} dv \int_0^{\phi(v)} \frac{\partial^2\sqrt{G}}{\partial u^2}\,du,$$

where $u = \phi(v)$ is the equation of the geodesic A_2A_3. Note that we are first to integrate with respect to u along the geodesic A_1B_1 (Fig. 33) between $u = 0$ and $u = \phi(v)$, and are then to integrate the result with respect to v from $v = 0$ to $v = \alpha_1$, that is, as the geodesic A_1B_1, starting from the position A_1A_2, sweeps out the triangle.

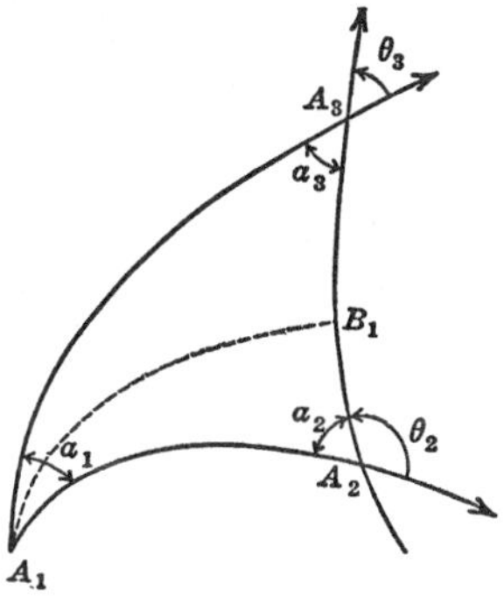

FIG. 33

The integration with respect to u gives

$$T = \int_0^{\alpha_1}\left(\frac{\partial\sqrt{G}}{\partial u}\right)_{u=0} dv - \int_0^{\alpha_1}\left(\frac{\partial\sqrt{G}}{\partial u}\right)_{u=\phi(v)} dv.$$

The integrand in the first of these integrals has, according to (6), the value unity. That in the second integral is equal, by (12), to $-d\theta/dv$, where θ is the angle under which the directed geodesic A_2A_3 cuts the u-curves. Hence

$$T = \int_0^{\alpha_1} dv + \int_{\theta_2}^{\theta_3} d\theta,$$

where θ_2 and θ_3 are the values of θ for $v = 0$ and $v = a_1$. Thus,

$$T = a_1 + \theta_3 - \theta_2.$$

It is evident that $\theta_2 = \pi - a_2$ and $\theta_3 = a_3$ (Fig. 33); hence

$$T = a_1 + a_2 + a_3 - \pi. \tag{15}$$

If K is always of one sign or always zero, the integral (13) is positive, zero, or negative according as $K > 0$, $K = 0$, or $K < 0$. Hence, the sum of the angles of a geodesic triangle is π, if the surface is a developable; greater than π, if the surface is of positive curvature; and, less than π, if the surface is of negative curvature.

THEOREM 2. *The difference from π of the sum of the angles of a geodesic triangle on a surface whose curvature is always of one sign is equal to the area of the triangle on the Gauss sphere which represents the given geodesic triangle.*

According to Chapter VI, (18), $\mathfrak{D} = \pm KD$, according as $K > 0$ or $K < 0$. Thus, the element of area, $d\mathcal{A} = \mathfrak{D}dudv$, on the sphere has the value $d\mathcal{A} = \pm KdA$, where dA is the corresponding element of area on the surface. The integral curvature T of a geodesic triangle on the surface is, then, equal to $\pm \mathcal{A}$, where $\mathcal{A}$ is the area of the corresponding triangle on the sphere. Hence

$$\mathcal{A} = a_1 + a_2 + a_3 - \pi \quad \text{or} \quad \mathcal{A} = \pi - a_1 - a_2 - a_3, \tag{16}$$

according as $K > 0$ or $K < 0$. Thus, the theorem is proved.

Area of a geodesic triangle on a surface of constant curvature. If K is a constant, not zero, (13) becomes

$$T = K\int\int dA = KA,$$

where A is the area of the given geodesic triangle. Consequently, we have, from (14),

$$A = \frac{a_1 + a_2 + a_3 - \pi}{K}. \tag{17}$$

The content of this equation may be stated as follows:

THEOREM 3. *The area of a geodesic triangle on a surface of constant total curvature, not zero, is equal to the excess over π or the deficit from π of the sum of the angles of the triangle, divided by the numerical value of the curvature, according as $K > 0$ or $K < 0$.*

In the case of a sphere of radius r, formula (17) reduces to the usual formula, $A = r^2(a_1 + a_2 + a_3 - \pi)$, for the area of a triangle formed by great circles on the sphere.

Non-Euclidean geometries. We have noted (§ 59) that a surface of constant curvature admits as many deformations into itself as there are rigid motions of the plane into itself. We know also that it can be mapped geodesically on the plane. On the other hand, we have just seen that the sum of the angles of a geodesic triangle on it is not, in general, equal to two right angles.

The explanation of these facts is connected with non-Euclidean geometry. The reader is perhaps familiar with the controversy over Euclid's parallel postulate and knows that, if this postulate is suppressed, the remaining postulates lead to three different types of geometry, the geometry of Euclid, the hyperbolic geometry of Lobatschewsky, and the elliptic geometry of Riemann. He may also be familiar with the fact that in these three geometries the sum of the angles of a triangle is, respectively, equal to π, less than π, and greater than π.

The absolute geometry of a surface of constant curvature is, in the small, a complete realization of a non-Euclidean geometry; that of a surface of positive curvature reproduces an elliptic geometry, and that of a surface of negative curvature, a hyperbolic geometry. For example, the absolute geometry of a sufficiently small region on a sphere is an elliptic geometry, and that of a restricted portion of the surface of revolution of the tractrix is a hyperbolic geometry.

65. Geodesic curvature as an absolute property. We are now in a position to establish the validity of a definition of geodesic curvature which is entirely analogous to that of the ordinary curvature of a plane curve.

THEOREM 1. *The geodesic curvature of a directed curve C at a point P is equal to the limit, when Δs approaches zero, of the ratio $\Delta\psi/\Delta s$, where $\Delta\psi$ is the directed angle from the geodesic tangent to C at P to the geodesic tangent to C at a neighboring point P', and Δs is the directed arc PP'.*

Take as the u-curves the geodesics which cut C orthogonally, and, as the v-curves, the orthogonal trajectories of these geodesics. Then, if the parameter u is chosen as the common arc of the geodesics, measured from the curve C, u and v are geodesic parameters and C is the curve $u = 0$.

We assume first that the geodesics tangent to C at P: $(0, v_0)$ and P': $(0, v_0 + \Delta v)$ are to the right of C, as C is traced in the positive direction. For the sake of definiteness, we think of v as increasing in the positive direction along C and take $\Delta v > 0$.

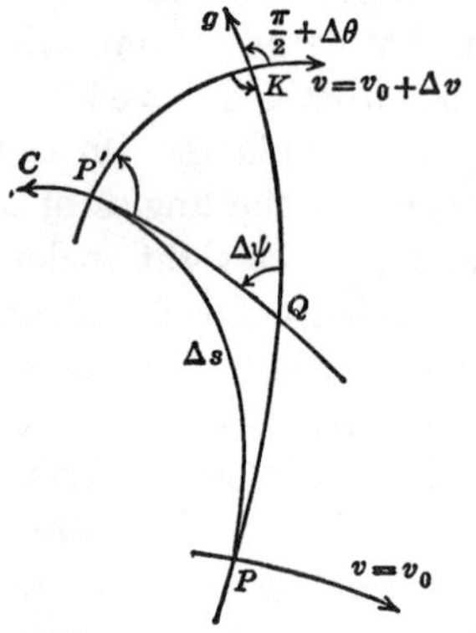

FIG. 34

We next consider the directed geodesic, g, tangent to C at P, and introduce the directed angle θ under which g cuts the u-curves. For $v = v_0$, that is, at P, $\theta = \pi/2$; and, for $v = v_0 + \Delta v$, that is, at the point K shown in Fig. 34, $\theta = \pi/2 + \Delta\theta$.

In the geodesic triangle QKP', the directed angles of the figure are all positive and are equal, respectively, to $\Delta\psi$, $\pi/2 + \Delta\theta$, and $\pi/2$. Since their sum is $\Delta\psi + \Delta\theta + \pi$, we have, by § 64, Theorem 1,

$$\Delta\psi + \Delta\theta = T,$$

where T is the integral curvature of the triangle QKP'. Hence

$$\lim_{\Delta s \to 0} \frac{\Delta\psi}{\Delta s} = -\left(\frac{d\theta}{ds}\right)_{\substack{u=0\\v=v_0}} + \lim_{\Delta s \to 0} \frac{T}{\Delta s}\cdot$$

According to the Gauss equation, (12), for the geodesic g, $d\theta = -(\partial\sqrt{G}/\partial u)dv$, and, therefore, since $ds = \sqrt{G}\,dv$,

$$\frac{d\theta}{ds} = -\frac{1}{\sqrt{G}}\frac{\partial\sqrt{G}}{\partial u} = -\frac{\partial \log \sqrt{G}}{\partial u}\cdot$$

On the other hand, inasmuch as the sides of the triangle QKP' all approach zero with Δs, T is an infinitesimal of higher order than Δs and so $T/\Delta s$ approaches zero. Thus

$$\lim_{\Delta s \to 0} \frac{\Delta\psi}{\Delta s} = \left(\frac{\partial \log \sqrt{G}}{\partial u}\right)_{\substack{u=0\\v=v_0}}\cdot$$

According to Chapter VII, (8), since $E = 1$, the right-hand member of this equation is the geodesic curvature, $1/\rho$, of the curve C at the point P. Hence the theorem is proved.

In the case which we have considered, namely, the case in which the concave side of C is on our left when we trace C in the positive sense, the geodesic curvature of C is positive. On the other hand, if the concave side of C is on the right when we trace C positively, the geodesic curvature of C is negative. For, in this case, it is clear from Fig. 35 that, when $\Delta s > 0$, then $\Delta\psi < 0$.

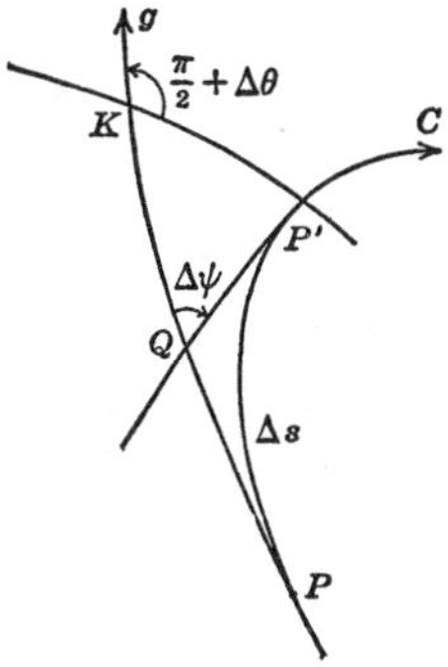

FIG. 35

The positive angles of the triangle QKP' in Fig. 35 are respectively $-\Delta\psi$, $\pi/2 - \Delta\theta$, and $\pi/2$; hence, the proof of the theorem, in this case, begins with the equation $-\Delta\psi - \Delta\theta = T$, and proceeds as before.

THEOREM 2. *The geodesic curvature of the curve C is equal numerically to the ordinary curvature of the plane curve into which C is deformed when the developable which is the envelope of the tangent planes to the surface at the points of C is rolled out on a plane.*

Inasmuch as the developable has the same tangent planes at the points of C as has the surface, C has the same geodesic curvature, as a curve on the developable, as it has as a curve on the surface. But the geodesic curvature of C, as a curve on the developable, is not changed when the developable is rolled out on a plane, and the geodesic curvature of the plane curve into which C is thereby deformed is equal numerically to its ordinary curvature.

COROLLARY. *The curve C is a geodesic if and only if, when the developable tangent to the surface along C is rolled out on a plane, the curve C, considered as a curve on the developable, is deformed into a straight line.*

66. The parallelism of Levi-Civita. As a finishing touch to the absolute geometry of a surface, we propose to develop a concept of parallelism on the surface which shall be as nearly as possible analogous to the ordinary concept of parallelism in the plane.

It is property of parallelism in the plane that, when a vector is carried parallel to itself so that its initial point traces a straight line—a geodesic of the plane—, the vector always makes the same angle with the straight line. Accordingly, we shall demand that parallelism on a surface be so defined that, when a vector tangent to the surface is carried parallel to itself so that its initial point traces a geodesic, the vector always makes the same angle with the geodesic. It is understood, of course, that the vector remains always tangent to the surface, that is, that its position at any point of the geodesic is in the tangent plane at the point.

The moving vector takes on the positions of infinitely many vectors, one at each point of the geodesic. When these vectors are thought of as attached to the geodesic as a curve of the developable which is tangent to the surface along the geodesic and this developable is rolled out on a plane, the geodesic becomes a straight line and, since angle is preserved, the vectors attached to it become vectors which issue from the points of the line and are actually parallel in the ordinary sense.

This relation of the displacement of a vector by parallelism along a geodesic to the ordinary parallel translation of a vector along a straight line in the plane suggests a suitable definition of parallel displacement of a vector along an arbitrary curve C on the surface.

Let V be a vector tangent to the surface at a point P and let C be a curve through P. Roll the developable which is tangent to the surface along C out on a plane. Thereby C becomes a plane curve $\bar{C}$ and the vector V becomes a vector $\bar{V}$ issuing from the point $\bar{P}$ of $\bar{C}$ corresponding to P. Translate $\bar{V}$ by ordinary parallelism so that $\bar{P}$ traces $\bar{C}$, thus obtaining a vector at each point of $\bar{C}$. Finally, wrap the plane about the surface in its original form and position. Thereby, the vectors at the points of $\bar{C}$ become tangent vectors to the surface at the points of C. It is these vectors which we define as the vectors into which the given vector V is carried by parallel displacement along the curve C.

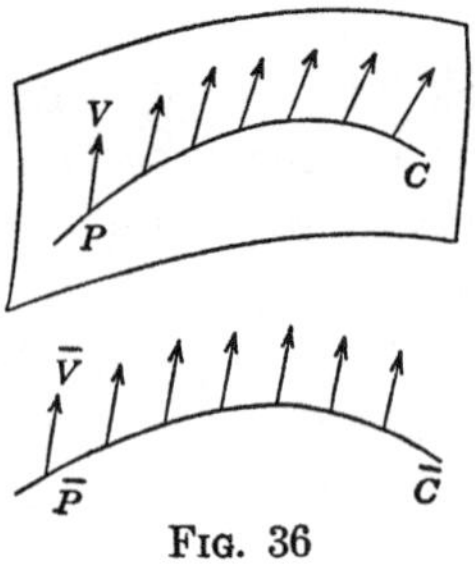

FIG. 36

This definition of parallel displacement of a vector along a curve has the advantage that it is readily visualized and comprehended. It has the disadvantage that it is not absolute, but depends on the surrounding space.*

The following theorem is a direct consequence of the definition which we have given.

THEOREM 1. *A necessary and sufficient condition that a curve be a geodesic is that a tangent vector to the curve, when it is displaced by parallelism along the curve, remain always tangent to the curve.*

We are now in a position to give new absolute interpretations of geodesic curvature and total curvature.

THEOREM 2. *Let P' be a point neighboring to a given point P of a directed curve C, V' the tangent vector to C at P', and V the vector at P' which results from the parallel displacement, along C, of the tangent vector to C at P. Then the geodesic curvature of C at P is equal to the limit, when Δs approaches zero, of $\Delta\psi/\Delta s$, where $\Delta\psi$ is the directed angle from V to V' and Δs is the directed arc PP'.*

THEOREM 3. *Let an arbitrary vector tangent to the surface be subjected to a parallel displacement so that its initial point makes a complete circuit of a closed curve in the direction which keeps the region bounded by the curve always to the left. If $\Delta\psi$ is the directed angle from the vector in its initial position to the vector in its final position, and ΔS is the area of the region, then the limit of $\Delta\psi/\Delta S$, when the closed curve shrinks to a point within the region, is the total curvature at the point.*

We illustrate the latter theorem in the case of the sphere. Let the closed curve be the circle of colatitude ϕ and let the initial position V_I of the vector be tangent to the meridian at the initial point P of the circuit. Slit the cone which is tangent to the sphere at the points of the circle along the ruling MP and roll it out on a plane (Fig. 38). Then translate the given vector parallel to itself along the arc of the circle in the plane which represents the circle on the sphere. The angle $\Delta\psi$ from the initial position V_I of the vector to the final position V_F is evidently equal to the smaller angle at M. Since the radius of the circle on the sphere

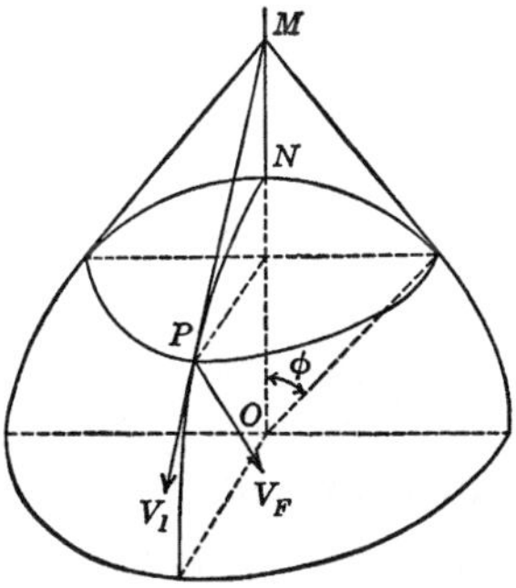

FIG. 37

* A definition which is absolute will be found in H. Weyl, *Space, Time, Matter*, § 14.

is $a \sin \phi$, where a is the radius of the sphere, the length of the circular arc in the plane is $2\pi a \sin \phi$. The radius MP of this arc is $a \tan \phi$. Hence $\Delta\psi = 2\pi - 2\pi \cos \phi$. The area ΔS of the smaller zone of the sphere bounded by the given circle is $2\pi a^2(1 - \cos \phi)$. Therefore $\Delta\psi/\Delta S = 1/a^2$ and lim $\Delta\psi/\Delta S$, when ϕ approaches zero, that is, when the circle shrinks to a point, is $1/a^2$. But, we know that $1/a^2$ is the total curvature of the sphere at every point.

Theorem 2 may be applied in a similar fashion to show that the geodesic curvature of the circle of colatitude ϕ is constant and equal to $(\cot \phi)/a$.

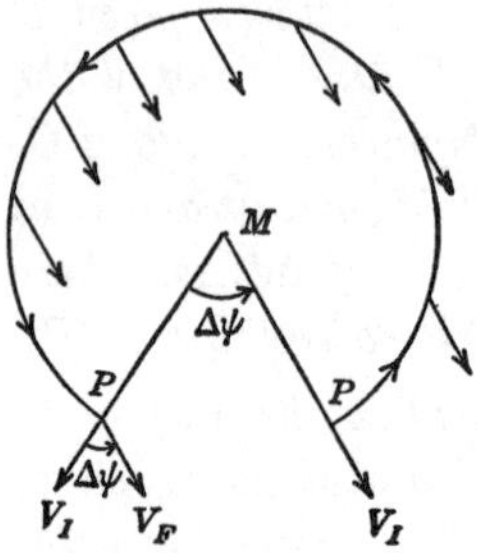

FIG. 38

The reader may have been surprised that we have failed to state conditions under which two vectors V and V' tangent to a surface at two distinct points P and P' are parallel. To expect such conditions to exist is to assume that no matter along what curve connecting P to P' the vector V is carried by parallelism, the resulting vector V' at P' is always the same. This would mean that when a vector V is carried parallel to itself around a closed circuit, it would return to its initial position. Hence, by Theorem 3, the curvature of the surface would have to be zero. It follows that, if the surface is not a developable or a plane, the parallelism of the two vectors V and V' at P and P' is dependent upon the path from P to P'. In other words, the parallelism is not absolute, but relative, with respect to a certain curve.

Even on a developable surface, parallelism, though absolute in the small, is not necessarily absolute in the large; for, it is evident that the vector in Fig. 37, considered as a vector on the cone, does not return to itself when it is carried by parallelism completely around the circle.

Analytic conditions for parallelism. Consider, with reference to the directed curve C: $u = u(s)$, $v = v(s)$ on a surface, a family of vectors, $\xi = \xi(s)$, tangent to the surface and issuing from the points of C. Let $\omega = \omega(s)$ be the directed angle, at an arbitrary point P of C, from the vector α tangent to C at P to the vector of the family at P. Then, if the positive direction for the measurement of ω is taken as the direction from the vector α to the vector ν of § 57, *the vectors of the family are mutually parallel with respect to C if and only if*

$$\frac{d\omega}{ds} + \frac{1}{\rho} = 0, \tag{18}$$

where $1/\rho$ is the geodesic curvature of C.

To prove the theorem, we make use of the results of § 57. Evidently, if $a(s)$ is the length of ξ,

$$\xi = a(\alpha \cos \omega + \nu \sin \omega).$$

Thus,

$$\frac{d\xi}{ds} = \frac{d \log a}{ds} \xi + a \left[\frac{d\alpha}{ds} \cos \omega + \frac{d\nu}{ds} \sin \omega + (-\alpha \sin \omega + \nu \cos \omega) \frac{d\omega}{ds} \right].$$

Substituting for $d\alpha/ds$ and $d\nu/ds$ the values found for them in § 57, and noting that

$$\eta = \widehat{\zeta\,\xi} = a(-\alpha \sin \omega + \nu \cos \omega),$$

we find that

$$\frac{d\xi}{ds} = \frac{d \log a}{ds} \xi + \left(\frac{d\omega}{ds} + \frac{1}{\rho} \right) \eta + a \left(\frac{\cos \omega}{r} - \frac{\sin \omega}{\tau} \right) \zeta . \tag{19}$$

According to the theorem of § 57, all the quantities in this equation remain unchanged when the curve C and the family of vectors $\xi = \xi(s)$ are referred to the developable which is tangent to the surface along C. Moreover, as this developable is rolled out on a plane, $1/\rho$, ω, $d\omega/ds$, a and da/ds remain the same. On the other hand, ξ, η, ζ and $1/r$, $1/\tau$ change, always satisfying, however, equation (19). In particular, $1/r$ and $1/\tau$ both approach zero. It follows, then, since the vectors ξ in the plane are mutually parallel if and only if $d\xi/ds$ is a multiple of ξ, that the vanishing of $d\omega/ds + 1/\rho$ is a necessary and sufficient condition that the given vectors ξ on the surface be mutually parallel with respect to C. Thus, the proposition is proved.

Condition (18) controls merely the direction of the vector ξ generating the given family. If, as is usually the case, it is required that the length a of ξ remain always the same, there should be added the condition $da/ds = 0$. But, according to (19), da/ds and $d\omega/ds + 1/\rho$ both vanish when and only when $d\xi/ds$ is a multiple of ζ. In other words:

THEOREM 4. *The vectors $\xi = \xi(s)$ tangent to a surface in the points of a curve C: $u = u(s)$, $v = v(s)$ all have the same length and are mutually parallel with respect to the curve C if and only if the vector $d\xi/ds$ is normal always to the surface.*

It should be noted that, when the vector generating the family is the unit vector α tangent to C, the theorem simply says that C is, by definition, a geodesic.

Since ξ is always tangent to the surface, we may write

$$\xi = \lambda_1 x_u + \lambda_2 x_v,$$

and think of the family of vectors as defined by the two functions λ_1, λ_2 of s. Then, *necessary and sufficient conditions that the vectors of the family all have the same length and are mutually parallel with respect to the curve C are*

$$(20)\quad \begin{aligned} \frac{d\lambda_1}{ds} + \left(C^1_{11}\frac{du}{ds} + C^1_{12}\frac{dv}{ds}\right)\lambda_1 + \left(C^1_{21}\frac{du}{ds} + C^1_{22}\frac{dv}{ds}\right)\lambda_2 = 0,\\ \frac{d\lambda_2}{ds} + \left(C^2_{11}\frac{du}{ds} + C^2_{12}\frac{dv}{ds}\right)\lambda_1 + \left(C^2_{21}\frac{du}{ds} + C^2_{22}\frac{dv}{ds}\right)\lambda_2 = 0, \end{aligned}$$

where $C^1_{21} = C^1_{12}$ and $C^2_{21} = C^2_{12}$.

For, the vector $d\xi/ds$ is normal always to the surface if and only if

$$\left(x_u\, x_v \middle| \frac{d\xi}{ds}\, x_v\right) = 0, \qquad \left(x_u\, x_v \middle| x_u\, \frac{d\xi}{ds}\right) = 0,$$

and when we set for $d\xi/ds$ its value, namely,

$$\frac{d\xi}{ds} = \frac{d\lambda_1}{ds}x_u + \frac{d\lambda_2}{ds}x_v + \lambda_1\left(\frac{du}{ds}x_{uu} + \frac{dv}{ds}x_{uv}\right) + \lambda_2\left(\frac{du}{ds}x_{uv} + \frac{dv}{ds}x_{vv}\right),$$

these equations take the above forms.

If the family consists of the unit vectors tangent to C, that is, if $\lambda_1 = du/ds$ and $\lambda_2 = dv/ds$, conditions (20) demand, as they should, that the curve C be a geodesic. For, they reduce precisely to equations (13) of Chapter VII.

67. The analytic theory in absolute form. Inasmuch as the absolute properties of a surface depend only on the linear element (§ 59) and may all be defined geometrically without leaving the surface, it devolves on us to express them analytically in terms of E, F, G and quantities definable on the surface, without reference to the surrounding space.

What we need here, primarily, are substitutes for the space components of a vector. More specifically, we need new components, relative to the curvilinear coordinate system on the surface, of a vector, V, on the surface, that is, a vector issuing from a point P: (u, v) of the surface and tangent to the surface. For this purpose, we introduce in the tangent plane at P oblique coordinates (λ_1, λ_2), referred to P as origin and to the directed tangents to the u- and v-curves through P as coordinate axes. In particular, we take as the coordinates (λ_1, λ_2) of a point Q (Fig. 39) the ratios of the directed distances $\overline{PQ_1}$ and $\overline{PQ_2}$ to $\sqrt{E}$ and $\sqrt{G}$:

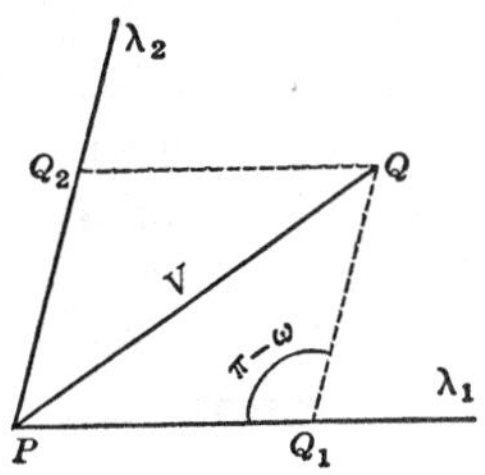

FIG. 39

$$\lambda_1 = \overline{PQ_1}/\sqrt{E}, \qquad \lambda_2 = \overline{PQ_2}/\sqrt{G}.$$

We then define the desired components of the vector V as the coordinates (λ_1, λ_2) of its terminal point Q and call them the surface components of V.

The relation between the surface components λ_1, λ_2 and the space components ξ_1, ξ_2, ξ_3 of the vector V is given by the symbolic equation

$$\xi = \lambda_1 x_u + \lambda_2 x_v. \tag{21}$$

For, since the unit vectors at P in the positive directions of the axes of λ_1 and λ_2 are $x_u/\sqrt{E}$ and $x_v/\sqrt{G}$ and $\overline{PQ_1} = \lambda_1\sqrt{E}$, $\overline{PQ_2} = \lambda_2\sqrt{G}$, we have $\xi = \lambda_1\sqrt{E}(x_u/\sqrt{E}) + \lambda_2\sqrt{G}(x_v/\sqrt{G}) = \lambda_1 x_u + \lambda_2 x_v$.

Comparison of (21) with Chapter IV, (19), shows that the ratio λ_2/λ_1 of the surface components of the vector V is precisely the direction coordinate, λ, of the line on which V lies.

Applying the law of cosines to the triangle PQ_1Q in Fig. 39 and recalling that $\sqrt{EG}\cos\omega = F$, we find that the square of the length of the vector V, expressed in terms of λ_1, λ_2, is

$$E\lambda_1^2 + 2F\lambda_1\lambda_2 + G\lambda_2^2.$$

The same result may be obtained directly by substituting the value of ξ given by (21) in $(\xi | \xi)$.

If θ is the angle between the two vectors at P with the components λ_1, λ_2 and μ_1, μ_2,

$$\cos\theta = \frac{E\lambda_1\mu_1 + F(\lambda_1\mu_2 + \lambda_2\mu_1) + G\lambda_2\mu_2}{\sqrt{E\lambda_1^2 + 2F\lambda_1\lambda_2 + G\lambda_2^2}\sqrt{E\mu_1^2 + 2F\mu_1\mu_2 + G\mu_2^2}}.$$

For, this is the result of setting $\xi = \lambda_1 x_u + \lambda_2 x_v$, $\eta = \mu_1 x_u + \mu_2 x_v$ in the usual formula for $\cos\theta$.

In passing judgment on the analytic form of the absolute geometry of a surface from the point of view above described, it will be instructive to write the fundamental formulas in terms of a new notation. Instead of u, v, we shall use u^1, u^2 and in place of E, F, and G, we shall employ g_{11}, g_{12} or g_{21}, and g_{22}. The linear element may, then, be written in the form

$$ds^2 = g_{11}du^1du^1 + g_{12}du^1du^2 + g_{21}du^2du^1 + g_{22}du^2du^2,$$

and hence expressed by the summation

$$(22) \qquad ds^2 = \Sigma g_{ij}du^idu^j,$$

where i and j range independently over 1, 2.

The expression for the square of the length of the vector at P: (u^1, u^2) with the surface components λ^1, λ^2 (instead of λ_1, λ_2) now reads

$$(23) \qquad \Sigma g_{ij}\lambda^i\lambda^j,$$

and the formula for the cosine of angle between the two vectors at P with the components λ^1, λ^2 and μ^1, μ^2 becomes

$$(24) \qquad \cos\theta = \frac{\Sigma g_{ij}\lambda^i\mu^j}{\sqrt{\Sigma g_{ij}\lambda^i\lambda^j}\sqrt{\Sigma g_{ij}\mu^i\mu^j}}.$$

In terms of the new notation, the differential equations of the geodesics—equations (13) of Chapter VII—, become

$$(25) \qquad \frac{d^2u^k}{ds^2} + \sum_{i,j} C^k_{ij}\frac{du^i}{ds}\frac{du^j}{ds} = 0, \qquad k = 1, 2,$$

and the differential equations, (20), for the parallelism of Levi-Civita take a similar form:

$$(26) \qquad \frac{d\lambda^k}{ds} + \sum_{i,j} C^k_{ij}\lambda^i\frac{du^j}{ds} = 0, \qquad k = 1, 2.$$

According to Chapter VI, (5), the C's are functions of the g's and their partial derivatives, and the quantities, other than the g's and C's, which enter into the foregoing equations, are all definable on the surface. Thus, we have achieved the desired end, an absolute form of the analytic theory of the absolute geometry of the surface.

We have not, however, divorced the geometry of the surface completely from the space in which the surface is imbedded. In fact, the very core of the geometry, namely, the measure of distance, belongs to the surrounding space. For, the linear element of the surface was obtained directly from the linear element, $(dx \mid dx)$, of the space, by substituting for x_1, x_2, x_3 the functions $x_1(u^1, u^2)$, $x_2(u^1, u^2)$, $x_3(u^1, u^2)$ which define the surface.

68. Riemannian geometry. Imagine, now, that the surrounding space is completely blotted out so that there is left merely the surface, referred to a system of curvilinear coordinates u^1, u^2. Choose arbitrarily, as the basis for measuring distance on the surface, a positive definite quadratic form, (22), whose coefficients are analytic functions of u^1, u^2. Then there is determined on the surface a geometry which is self-contained and finds its analytic expression in the formulas of § 67.

This is the geometry of Riemann, developed in his famous Habilitationsschrift, *Über die Hypothesen, welche der Geometrie zu Grunde liegen*, 1854. In describing it in greater detail, we agree to think of an ordered number pair λ^1, λ^2 as representing a vector V at the point P: (u^1, u^2) of the surface and to call λ^1, λ^2 the components of V. We then define the length of V by (23), and the angle formed by V and a second vector at P, with the components μ^1, μ^2, by (24).

The geodesics on the surface are obtained as the curves of shortest distance. The length of arc of the curve C: $u^1 = u^1(t)$, $u^2 = u^2(t)$ from the point P_1: $t = t_1$ to the point P_2: $t = t_2$ is defined by the integral

$$s = \int_{t_1}^{t_2} \sqrt{\Sigma g_{ij} \frac{du^i}{dt} \frac{du^j}{dt}}\, dt,$$

and the necessary conditions of Euler that this arc be shorter than any other analytic arc lying near to it and joining P_1 to P_2 actually reduce, when the arc s of C is introduced in place of the parameter t, to equations (25).

The parallelism of Levi-Civita may be based on the definition of Weyl (§ 66). More often, however, it is defined outright by formulas (26). The (geodesic) curvature of a curve and the curvature of the surface are then obtainable by the procedures described in Theorems 2 and 3 of § 66.

The foregoing description of the Riemannian geometry of a surface, or two-dimensional manifold, admits immediate extension to the case of an n-dimensional manifold. Here, a point has n coordinates $u^1, u^2, \cdots, u^n$, a vector has n components $\lambda^1, \lambda^2, \cdots, \lambda^n$, and the assumed linear element has n^2 coefficients g_{ij}, not all distinct, of course, since always $g_{ij} = g_{ji}$. Imagine, then, that the summations in the preceding formulas, and the superscripts i in (25) and (26), range over $1, 2, \cdots, n$. Then, the formulas present the fundamentals of the analytic theory in the n-dimensional case.

There is, however, one difficulty. It is not obvious how to generalize the Christoffel symbols, as given in Chapter VI, (5). But, these symbols are expressible in a form capable of immediate generalization, namely,

$$C_{ij}^k = \frac{1}{2}\sum_h g^{kh}\left(\frac{\partial g_{jh}}{\partial u^i} + \frac{\partial g_{ih}}{\partial u^j} - \frac{\partial g_{ij}}{\partial u^h}\right).$$

Here, $g^{ij} = G_{ij}/g$, where g is the determinant, $|g_{ij}|$, in which the element in the i-th row and j-th column is g_{ij}, and G_{ij} is the cofactor of g_{ij} in this determinant.

The notation introduced in § 67 gives a slight idea of the general analytic method, known as the absolute differential calculus or tensor analysis, which is now universally employed in the treatment of Riemannian geometry. In essence, this method goes back to Riemann's *Habilitationsschrift* and the work of Christoffel (1869). The credit for it as a complete mathematical discipline is due to Ricci (1892). But it was not until later that the method and, for that matter, Riemannian geometry itself, received proper recognition, through their importance in the theory of relativity.

Non-Riemannian geometry is essentially an outgrowth of Levi-Civita's concept of parallelism (1917). Like the geometry of Riemann, it deals with an n-dimensional manifold of points $(u^1, u^2, \cdots, u^n)$. But it has nothing to do with distance. Instead, it is based either on a concept of displacement, a generalization of Levi-Civita's parallelism, or on a concept of paths, a generalization of the idea of geodesics. In the *geometry of linear displacement,* inaugurated by Weyl (1918), the point of departure is the system of differential equations (26) for the displacement of a vector along a curve, and, in the *geometry of paths,* founded by Eisenhart and Veblen (1922), it is the system of differential equations (25) which is fundamental. In both cases, the quantities C^k_{ij} are assumed outright, as functions of $u^1, u^2, \cdots, u^n$.

CHAPTER X

SURFACES OF SPECIAL TYPE

69. Surfaces of revolution. In § 61 we learned that the equations

$$(1) \qquad x_1 = r\cos v, \quad x_2 = r\sin v, \quad x_3 = f(r), \qquad r \geqq 0,$$

represent a surface of revolution referred to the x_3-axis as axis and to the meridians and parallels as parametric curves.

The fundamental quantities for the surface are found to be:

$$(2) \qquad \begin{aligned} E &= 1 + f'^2, \quad F = 0, \quad G = r^2, \\ e &= \frac{f''}{\sqrt{1+f'^2}}, \quad f = 0, \quad g = \frac{rf'}{\sqrt{1+f'^2}}. \end{aligned}$$

It follows from these formulas or, more simply, from the geometric theory of § 44, that the lines of curvature are the meridians and the parallels, and that the radii of principal normal curvature ($r_1 = E/e$ and $r_2 = G/g$) at a point P are equal respectively, except perhaps for sign, to the radius of curvature at P of the meridian passing through P and the distance from P to the axis measured along the normal to the surface at P.

The only minimal surface of revolution is the catenoid. For, $K' = 0$ only if $Eg + Ge = 0$ or $f'(1 + f'^2) + rf'' = 0$, and this differential equation can be rewritten in the form

$$\frac{df'}{f'(1+f'^2)} + \frac{dr}{r} = 0,$$

and hence has the first integral

$$\frac{f'r}{\sqrt{1+f'^2}} = c, \quad \text{or} \quad df = \frac{c\,dr}{\sqrt{r^2 - c^2}}.$$

Then $f(r) = c\cosh^{-1}(r/c) + k$ and the curve $x_3 = f(r)$ is a catenary with the axis of x_3 as axis.

The differential equation of the asymptotic lines of the general surface of revolution is, according to (2), $f''dr^2 + rf'dv^2 = 0$. Hence, the finite equations are of the form $v = \pm \int \psi(r)dr + c$. We express this result by saying that *the asymptotic lines may be found by a quadrature*, that is, by an integration.

Inasmuch as $F = 0$ and E/G is a function of r alone, *the meridians and parallels form an isometric system.* Corresponding isometric parameters are u, v, where

$$u = \int \frac{1}{r}\sqrt{1 + f'^2}\, dr. \tag{3}$$

For, when we set $(1 + f'^2)dr^2 = r^2du^2$, the linear element evidently becomes

$$ds^2 = r^2(du^2 + dv^2), \tag{4}$$

where r is the function of u which is the inverse of the function $u = \phi(r)$ defined by (3).

Consider, now, the map of the surface on the plane $x_1 = u$, $x_2 = v$, $x_3 = 0$ in which corresponding points have the same curvilinear coordinates. This map orders to the meridians and parallels on the surface the lines parallel to the axes of coordinates in the plane. Furthermore, since the linear element of the plane is $ds^2 = du^2 + dv^2$, the map is conformal.

THEOREM 1. *A surface of revolution can be mapped conformally on a plane so that the meridians and parallels correspond to an orthogonal system of straight lines.*

The loxodromes on the surface, since they are by definition the curves cutting the meridians under constant angles, correspond to the straight lines in the plane. Hence, the finite equation of the loxodromes is $au + bv + c = 0$.

Mercator map of the earth. The sphere

$$x_1 = a \sin \phi \cos v, \quad x_2 = a \sin \phi \sin v, \quad x_3 = a \cos \phi \tag{5}$$

is the surface (1) for which $r = a \sin \phi$, $f(r) = a \cos \phi$. Substituting these values of r and $f(r)$ in the integral in (3), we obtain as a value of the integral $u = \log \tan \phi/2$. Hence, the equations $x_1 = \log \tan \phi/2$, $x_2 = v$, in conjunction with equations (5), establish a conformal map of the sphere on the plane in which the meridians and parallels correspond to straight lines.

Deformation of surfaces of revolution. In § 61, we noted that the linear element of (1) is reducible to the form

$$ds^2 = du^2 + r^2 dv^2, \tag{6}$$

where u is a function of r defined by the equation

$$du^2 = dr^2 + df^2, \tag{7}$$

and $r = U(u)$ is the inverse of this function.

These new parameters u, v are respectively the common arc of the meridian curves, measured from a particular parallel, and the angle of longitude. The parametric equations of the surface in terms of them are

$$x_1 = U \cos v, \quad x_2 = U \sin v, \quad x_3 = \int \sqrt{1 - U'^2}\, du. \tag{8}$$

For, when we set $r = U(u)$ in (7), we get $df^2 = (1 - U'^2)du^2$.

The linear element (6) is now $ds^2 = du^2 + U^2 dv^2$. It evidently remains unchanged when we replace U by cU and v by v/c, where c is any positive constant. Hence, all the surfaces

$$x_1 = cU \cos \frac{v}{c}, \quad x_2 = cU \sin \frac{v}{c}, \quad x_3 = \int \sqrt{1 - c^2 U'^2}\, du, \quad c > 0, \tag{9}$$

have the linear element

$$ds^2 = du^2 + U^2 dv^2 \tag{10}$$

in common with the surface (8). But these surfaces are all surfaces of revolution and form a continuous family. In other words:

THEOREM 2. *A surface of revolution can be deformed continuously into ∞^1 surfaces of revolution.*

Surfaces applicable to a surface of revolution. We are now in a position to prove Theorem 6 of § 61 to the effect that every surface which admits a continuous deformation into itself is applicable to a surface of revolution. According to § 61, Theorem 3, such a surface is characterized by the fact that its linear element is reducible to the form (10). It suffices then to show that there exists a surface of revolution with this linear element. The surface (8) apparently fulfils the requirement and the proof thus appears to be complete.

There is, however, a difficulty. In the foregoing work, the surface of revolution is given, so that we know that (8) is a real surface. Here, it is the positive function $U(u)$ which is given, and there is no guarantee that $1 - U'^2 > 0$, that is, that (8) is a real surface.

Instead of the surface (8) it is clear that we may employ any one of the surfaces (9) and herein lies the solution of the difficulty. Inasmuch as we are concerned only with a proof of the theorem in the small, we may restrict the consideration of $U(u)$ to a closed interval $a \leqq u \leqq b$. Since U is in any case analytic, U' has a maximum absolute value in this interval and hence c can be chosen so that throughout the interval $1 - c^2U'^2 > 0$. The corresponding surface (9) is then real and the proof is complete.

EXERCISES

1. Show that, if the sides of a triangle on a surface of revolution are arcs of loxodromes, the sum of the angles is equal to two right angles.

2. Prove that the finite equation of the geodesics on a surface of revolution can be found by a quadrature.

3. Show that a curve C on a surface of revolution, other than a parallel, is a geodesic if and only if $r \sin \theta$, where r is the radius of the parallel through a point P of C and θ is the angle at P under which C cuts the meridian through P, is constant along C. Hence prove that a curve, other than a meridian and a certain type of parallel, cannot be both a geodesic and a loxodrome unless the surface of revolution is a cylinder.

4. Prove that the map of the surface (1) on the plane,

$$x_1 = \int r\sqrt{1 + f'^2}\, dr, \quad x_2 = v, \quad x_3 = 0,$$

in which corresponding points have the same curvilinear coordinates, is area-preserving. Discuss the map in greater detail, first in the general case, and then in the case of the sphere.

5. Discuss the map of the surface (8) on the surface (9) when $c < 1$; when $c > 1$.

6. Prove that the differential equation of the tractrix in Fig. 28 is

$$r\, dx_3 + \sqrt{a^2 - r^2}\, dr = 0.$$

Hence show that, when $U = e^{-u/a}$, equations (8) represent the surface generated by the rotation of this tractrix about its asymptote, and that these equations can be reduced by a suitable change of parameters to the form

$$x_1 = a \sin \phi \cos v, \quad x_2 = a \sin \phi \sin v, \quad x_3 = -a(\cos \phi + \log \tan \phi/2).$$

7. Show that the ∞^1 surfaces of revolution into which the surface of revolution of a tractrix is continuously deformable are all identical with this surface. What are the path curves of the resulting continuous deformation of the surface into itself? *Ans.* The curves $v \sin \phi =$ const.

8. If the coefficients in the linear element of a surface are functions of u alone, the surface is applicable to a surface of revolution.

Helicoids. A surface generated by the screw motion of a curve about a fixed line, that is, a rigid motion such that the points of the curve trace circular helices (or circles) which lie on circular cylinders with the fixed line as axis, is called a *helicoid*. The fixed line is known as the *axis* and the sections by planes through it are called the *meridians*. Evidently, the meridians are congruent plane curves and the surface can equally well be generated by the proper screw motion of any one of them about the axis.

9. Show that the helicoid generated by the screw motion of the curve $x_3 = f(r)$ about the x_3-axis has parametric equations of the form

$$x_1 = r \cos v, \quad x_2 = r \sin v, \quad x_3 = f(r) + av.$$

Prove that the meridians and the helices form an orthogonal system only when the helicoid is a right helicoid or a surface of revolution.

10. The helices on any helicoid form an isometric family of geodesic parallels and their orthogonal trajectories can be found by a quadrature. Establish these facts and state the conclusions that may be drawn from them.

11. Prove that there are ∞^2 noncongruent helicoids which are minimal surfaces and that they are given by

$$f(r) = c \int \sqrt{\frac{r^2 + a^2}{r^2 - c^2}} \frac{dr}{r},$$

where a, c are arbitrary constants.

70. Ruled surfaces. A surface, other than a plane, which contains a family of straight lines is called a ruled surface and the straight lines, its rulings or generators. If the straight lines are imaginary, the surface, since it is assumed real, must also contain the conjugate-imaginary lines and hence must be a quadric surface with imaginary generators. Thus, when we exclude these quadrics, the rulings of the surface must be real.

The surface has the symbolic representation

$$y = x(v) + u\eta(v), \tag{11}$$

where $x = x(v)$ represents an arbitrarily chosen, but fixed, trajectory C of the rulings, $\eta(v)$ is a unit vector in the direction of the ruling L which passes through the point P of C whose coordinate is v, and u is the directed distance from P to an arbitrary point on L.

As the parameter v we take the arc of the curve $z = \eta(v)$ on the unit sphere. This curve is known as the *spherical indicatrix of the rulings* of the surface. In taking its arc as v, we are excluding cylinders from consideration, inasmuch as in the case of a cylinder the locus $z = \eta$ is not a curve, but a point.

It is to be noted that the arc of the spherical indicatrix is an invariant of the surface and hence is a better choice for v than the arc of the so-called *directrix* C, whose choice is arbitrary.

Since η is a unit vector and v is the arc of the spherical indicatrix,

$$(12)\quad (\eta\,|\,\eta) = 1, \quad (\eta\,|\,\eta') = 0, \quad (\eta'\,|\,\eta') = 1, \quad (\eta'\,|\,\eta'') = 0.$$

From $y_u = \eta$, $y_v = x' + u\eta'$, $y_{uu} = 0$ and $y_{uv} = \eta'$, we readily find that

$$(13)\quad \begin{aligned} &E = 1, \quad F = (x'\,|\,\eta), \quad G = (x'\,|\,x') + 2\,u(x'\,|\,\eta') + u^2, \\ &e = 0, \quad f = -\frac{1}{D}(x'\,\eta\,\eta'), \quad K = -\frac{(x'\,\eta\,\eta')^2}{D^4}. \end{aligned}$$

From the value of K we conclude that *a necessary and sufficient condition that the surface be a developable surface is that* $(x'\,\eta\,\eta') = 0$.

Henceforth, we exclude developable surfaces. The curvature of the surface is then, in general, negative.

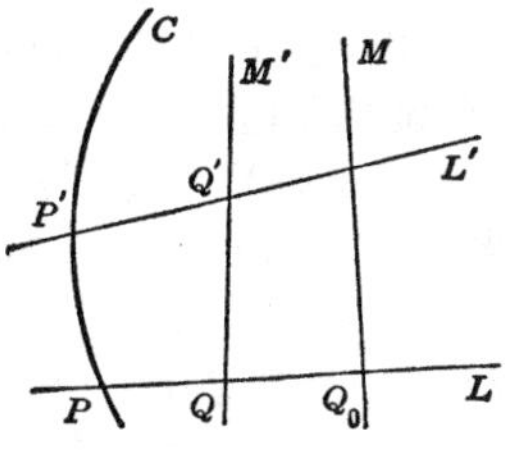

FIG. 40

Common perpendicular to two neighboring rulings. We shall have to do here with a point P' on C neighboring to P, the ruling L' through P', the line M' perpendicular to both L and L', and the line M which is the limit of M' when P' approaches P along C.

As we pass from P to P' along C, v takes on an increment Δv, and η the corresponding increment $\Delta\eta$. Direction cosines of L and L' are given, then, by η and $\eta + \Delta\eta$. Hence M' has the direction of the vector $\widehat{\eta\,\eta + \Delta\eta}$ or $\widehat{\eta\,\Delta\eta}/\Delta v$, and M that of the vector $\widehat{\eta\,\eta'}$. By (12), this vector is of unit length. Thus, *direction cosines of M are λ_1, λ_2, λ_3, where*

$$(14)\qquad \lambda = \widehat{\eta\,\eta'}$$

We seek next the coordinates of the point Q' (Fig. 40) in terms of those of Q and Δv. If Q: (u, v) has the symbolic coordinates y, those of any point on L' are $y + \Delta y$, where Δy is given by the Taylor's series

$$\Delta y = y_u \Delta u + y_v \Delta v + \cdots = \eta \Delta u + (x' + u\eta')\Delta v + \cdots.$$

This point will be the point Q' if the line joining it to Q is perpendicular to L, that is, if $(\Delta y | \eta) = 0$ or

$$\Delta u + (x' | \eta)\Delta v + \cdots = 0.$$

Consequently, for the point Q', Δu is equal to $-(x' | \eta)\Delta v$ to within terms of higher order in Δv and

$$\Delta y = (x' + u\eta' - (x' | \eta)\eta)\Delta v + \sigma, \tag{15}$$

where σ represents terms of at least the second order in Δv.

Since the triple $\Delta y/\overline{QQ'}$ constitutes direction cosines of M', its limit when Δv approaches zero is $\pm \lambda$. Hence $\Delta y = h(\lambda + \epsilon)$, where h is the distance QQ' properly directed and ϵ is a triple of infinitesimals. We may, then, replace (15) by

$$h(\lambda + \epsilon) = (x' + u\eta' - (x' | \eta)\eta)\Delta v + \sigma. \tag{16}$$

We are now in a position to discuss the common perpendicular distance, $|h|$, between L and L'. Taking the inner product of both sides of (16) with λ, dividing the resulting equation, $h(1 + (\epsilon | \lambda)) = (x' | \lambda)\Delta v + (\sigma | \lambda)$, through by Δv, and letting Δv approach zero, we find, since ϵ and $\sigma/\Delta v$ tend to zero, that

$$\lim_{\Delta v \to 0} \frac{h}{\Delta v} = (x' | \lambda) = (x' \eta\, \eta'). \tag{17}$$

Thus, *the distance between L and L′ is, in general, an infinitesimal of the first order with respect to* Δv.

The coefficient of Δv in (16) is evaluated at Q. Its limit, then, when Δv approaches zero, will be its value at the point Q_0 which is the limit of Q. With this in mind, we form the inner product of both sides of (16) with η', divide the resulting equation through by Δv, and take the limit when Δv approaches zero. The result is: $(x' | \eta') + u = 0$. Hence, *the coordinate u of the point* Q_0 *on L which is the limit of the foot of the common perpendicular to L and L′ is*

$$u = -(x' | \eta'). \tag{18}$$

Parameter of distribution. For reasons which will be evident presently, the point Q_0 is known as the *central point* of the ruling L and the tangent plane to it as the *central plane*. The locus of the central point, defined by equation (18) when v varies, is called the *line of striction.*

When a point Q traces a ruling of a developable surface, the tangent plane at Q remains the same. On the other hand, in the case of the general ruled surface, the tangent plane turns about the ruling. The law controlling its motion may be formulated as follows:

THEOREM 1. *The tangent of the directed angle from the central plane of a ruling L to the tangent plane at an arbitrary point Q of L is proportional to the directed distance from the central point Q_0 to the point Q.*

We shall measure the angle ϕ from the tangent plane at Q_0 to that at Q in the sense which is clockwise when we look along L in the positive direction of η. It follows, then, from § 36, that $\sin\phi = (\zeta_0\ \zeta\ \eta)$, where ζ_0 and ζ are respectively the normal vectors to the surface at Q_0 and Q. Since $\zeta_0 = \widehat{\eta\, x' + u_0\eta'}/D_0$ and $\zeta = \widehat{\eta\, x' + u\eta'}/D$, where u_0 and D_0 are evaluated at Q_0, we have

$$DD_0 \sin\phi = (\widehat{\eta\, x'} + u_0\widehat{\eta\,\eta'}\ \ \widehat{\eta\, x'} + u\widehat{\eta\,\eta'}\ \ \eta) = (\widehat{\eta\, x'}\ \widehat{\eta\,\eta'}\ \eta)(u - u_0).$$

But

$$(\widehat{\eta\, x'}\ \widehat{\eta\,\eta'}\ \eta) = (\eta\, x' | \widehat{\eta\,\eta'}\ \eta) = -\,(x'\ \eta\ \eta'),$$

and therefore

$$DD_0 \sin\phi = -\,(x'\ \eta\ \eta')(u - u_0).$$

In computing $\cos\phi$, from $\cos\phi = (\zeta|\zeta_0)$, we first rewrite the expression for ζ_0 by introducing for u_0 its value from (18). Since $(\eta'|\eta') = 1$, $x' + u_0\eta'$ has the value $(\eta'|\eta')x' - (x'|\eta')\eta'$ or $\widehat{\eta'\,\widehat{x'\,\eta'}}$. Hence $D_0\zeta_0 = \widehat{\eta\,\xi}$, where $\xi = \widehat{\eta'\,\widehat{x'\,\eta'}}$. Writing $D\zeta = \chi$, where $\chi = \widehat{\eta\, x' + u\eta'}$, we have, then,

$$DD_0 \cos\phi = (\chi|\widehat{\eta\,\xi}) = (\widehat{\chi\,\eta}|\xi) = (\chi\,\eta|\widehat{\eta'\,x'}\,\eta').$$

But

$$\begin{aligned}(\chi\,\eta|\widehat{\eta'\,x'}\,\eta') &= -\,(\eta\ \eta'\ x')(\chi|\eta') \\ &= -\,(x'\ \eta\ \eta')(\eta\ x' + u\eta'\ \eta') = (x'\ \eta\ \eta')^2,\end{aligned}$$

and hence

$$DD_0 \cos\phi = (x'\ \eta\ \eta')^2.$$

It follows now that

$$\tan \phi = \frac{u - u_0}{b}, \tag{19}$$

where

$$b = -(x' \, \eta \, \eta'). \tag{20}$$

and the theorem is established.

The function b is an invariant of the surface, according to (19). It is known as the *parameter of distribution.*

Returning to (17), we note that Δv is the arc of the spherical indicatrix between the points which represent the rulings L and L' and hence infer that the limit of the ratio of Δv to the angle between L and L' is ± 1. Thus, equation (17), in conjunction with (20), gives the following new interpretation of b.

THEOREM 2. *The limit of the ratio of the distance between L and L′ to the angle between L and L′, when L′ approaches L, is equal to the numerical value of the parameter of distribution for the ruling L.*

From (13) and (20), it follows that $K = -b^2/D^4$. To evaluate D, we assume for the moment that the directrix C is the line of striction, that is, that $(x' | \eta') = 0$. Then D^2 is equal to $(x' | x') \sin^2 \theta + u^2$, where $\sin \theta$ is the angle between the vectors x' and η. But η' is now perpendicular to both x' and η and hence, by § 36, $(x' \, \eta \, \eta')^2$ or b^2 has the value $(x' | x') \sin^2 \theta$. Thus, when u is measured from the line of striction, $D^2 = u^2 + b^2$. Consequently, when the directrix is arbitrarily chosen,

$$D^2 = (u - u_0)^2 + b^2, \tag{21}$$

and hence

$$K = -\frac{b^2}{[(u - u_0)^2 + b^2]^2}. \tag{22}$$

THEOREM 3. *The total curvature is the same at two points of a ruling which are equidistant from the central point and has its maximum numerical value at the central point.*

Suppose, finally, that the directrix C is an orthogonal trajectory of the rulings. Then $F = (x' | \eta) = 0$ and all the curves $u =$ const. meet the rulings at right angles. Furthermore,

$G = D^2$ and the linear element takes the simple form

$$ds^2 = du^2 + [(u - u_0)^2 + b^2]dv^2. \tag{23}$$

EXERCISES

1. Prove that the points of a ruling other than the central point may be paired so that the tangent planes at the points of each pair are perpendicular. Show that the points of a pair are on opposite sides of the central point and that the product of the distances from them to the central point is the same for all pairs.

2. Prove that $\sin \theta(ds/dv) = \pm b$, where s is the arc of the line of striction, v the arc of the spherical indicatrix, and θ is the angle under which the generic ruling meets the line of striction.

3. Show that the equations $x_1 = u \cos v$, $x_2 = u \sin v$, $x_3 = av$ of the right helicoid are of the form (11). Hence, reinterpret the results of § 42, proving that the line of striction of the helicoid is the axis and that the parameter of distribution is constant. What is the spherical indicatrix of the rulings?

4. Find the line of striction and the parameter of distribution of the right conoid. Show that, if the parameter of distribution is constant, or if each two secondary asymptotic lines cut equal segments from the rulings, the conoid is a helicoid.

5. A ruled surface one of whose secondary asymptotic lines is a straight line is given. Show that this straight line is the line of striction when and only when the spherical indicatrix of the rulings lies in a plane which is perpendicular to the line.

6. Prove that on a hyperboloid of revolution of one sheet the parallel of least radius is the line of striction, the rulings cut it at a constant angle, and the parameter of distribution is constant.

7. Prove geometrically that the right helicoid is the only ruled surface which is minimal.

8. Show that a cross-ratio of four tangent planes to a ruled surface at points of a ruling is equal to the corresponding cross-ratio of the points.

9. Show that the normals to a ruled surface in the points of a ruling form a hyperbolic paraboloid.

10. Show that for the surface (11)

$$Dg = (x' x'' \eta) + u[(x'' \eta \eta') - (x' \eta \eta'')] + u^2(\eta \eta' \eta'').$$

11. Prove that each two secondary asymptotic lines of a ruled surface cut equal segments from the rulings when and only when the rulings are parallel to a plane and the parameter of distribution is constant. First show that the surface has the desired property if and only if, when it is referred to its asymptotic lines as the parametric curves, it admits a representation of the form (11).

12. Prove analytically that the only minimal ruled surface is the right helicoid.

71. Translation surfaces. If C and C' are two space curves with a point P_0 in common, the translation of the curve C so that the point of C originally at P_0 traces the curve C' generates a surface, and the translation of the curve C' so that the point of C' originally at P_0 traces the curve C generates a second surface. These two surfaces are identical, as we shall show presently. The single surface is known as a *translation surface* and the curves into which C and C' are carried by the translatory motions are called its *generators*.

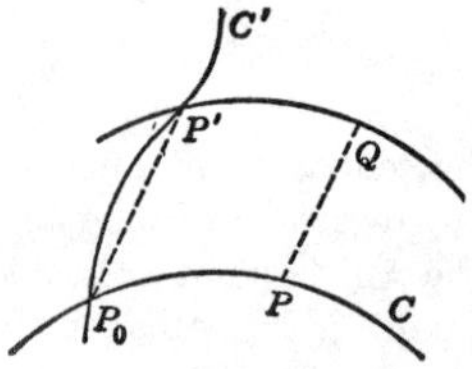

FIG. 41

A careful choice of the parametric representations of the curves C and C' will facilitate the proof that the two surfaces in question are the same, and also the subsequent work. We start with the representations $y = y(u)$, $z = z(v)$, assuming that $y(u_0) = z(v_0) = c$, that is, that the point P_0 has the coordinate u_0 on C, the coordinate v_0 on C', and the space coordinates (c_1, c_2, c_3). We next choose arbitrarily two triples of constants a and b whose sum is the triple c, and set $y(u) - b = U(u)$ and $z(v) - a = V(v)$. Putting $u = u_0$ and $v = v_0$ in these relations, we find that $a = U(u_0)$, and $b = V(v_0)$. Hence we obtain parametric representations of C and C',

$$(24) \qquad y = U(u) + V(v_0), \qquad z = U(u_0) + V(v),$$

which fulfil automatically the condition that the point $u = u_0$ of C and the point $v = v_0$ of C' be identical.

If P' is a point tracing C' and P a point tracing C, the locus of the point Q into which P is carried when C is translated along C' into the position through P' is the first of the surfaces described. The coordinates (x_1, x_2, x_3) of Q are evidently obtainable from the coordinates of P by adding the components of the vector $\overline{PQ}$ or those of the equal vector $\overline{P_0P'}$. But P has the symbolic coordinates $U(u) + V(v_0)$ and the vector $\overline{P_0P'}$, the symbolic components $U(u_0) + V(v) - U(u_0) - V(v_0)$ or $V(v) - V(v_0)$. Hence, a parametric representation of the first surface is

$$(25) \qquad x = U(u) + V(v).$$

In the generation of the second surface, the rôles of the curves C and C' are interchanged. This amounts, according to (24), to interchanging $U(u)$ and $V(v)$, and u_0 and v_0. Thereby, (25) is unchanged, and so the second surface is identical with the first.

Our translation surface is, then, represented by the symbolic equation (25). The generators are the parametric curves. In particular, the generators through the point (u_0, v_0) are the original curves (24). Since (25) is independent of u_0, v_0, these curves might have been any two generators of different families. Hence a translation surface is obtainable by the translation of any generator of the one family along any generator of the other.

THEOREM 1. *A necessary and sufficient condition that a surface be a translation surface is that it admit a parametric representation of the form* (25).

For, we have shown that every translation surface admits a representation of the desired type, and the converse is obvious.

THEOREM 2. *A translation surface is, in infinitely many ways, the locus of the mid-points of the line-segments joining the points of one curve to those of a second, and conversely.*

For, we may rewrite (25) in the form

$$x = \tfrac{1}{2}(Y + Z), \quad \text{where} \quad Y = 2\,U + c, \quad Z = 2\,V - c,$$

and c is an arbitrary triple of constants. Incidentally, we note that the two curves have respectively the same shapes as C and C', but are of twice the size.

THEOREM 3. *The generators of a translation surface, other than a developable or a plane, form a conjugate system.*

From (1), $x_{uv} = 0$ and so $f = 0$. Hence, the conclusion follows.

An interesting kind of translation surface results when the curves C and C' are the same curve. The parameters u and v, since they are always arbitrary, may be taken, in this case, as the same parameter, except for notation. Then U_1, U_2, U_3 are the same functions of u as V_1, V_2, V_3 are of v, and $u_0 = v_0$. Consequently, for *every* pair of equal values of u_0 and v_0 the two generators (24) are the same, and the two families of generators coincide in a single family. Through each point (u_0, v_0) of the surface for which $u_0 \neq v_0$ there pass two generators of this family, whereas through a point (u_0, v_0) for which $u_0 = v_0$ there passes only one. It follows, then, that the curves of the family cover all the points of the surface twice except those of the curve

$u = v$ and that every point other than one on this curve has two sets of curvilinear coordinates of the forms (a, b) and (b, a).

The curve $u = v$ has the parametric representation $X = 2\,U$ or $Y = 2\,V$, and hence the mid-points of its chords are the points of the surface itself. Thus a translation surface of the kind in question is characterized by the property that it is the locus of the mid-points of the chords of a curve which lies on it. The generators of the surface are similar to this curve, but of half the size.

Translation surfaces with imaginary generators. By an imaginary, or better a complex, analytic curve we mean the locus of complex points, that is, real and imaginary points, represented by the symbolic equation $x = x(t)$, where $x_1(t)$, $x_2(t)$, $x_3(t)$ are analytic functions, not all constant, of the complex variable $t = r + si$. By a complex analytic surface we mean the locus of complex points represented by a symbolic equation $x = x(u, v)$, where $x_1(u, v)$, $x_2(u, v)$, $x_3(u, v)$ are analytic functions of the two complex variables u, v such that the vector $x_u \frown x_v$ formed in the usual way is not identically the null vector. A complex curve consists of ∞^2 points, corresponding to the ∞^2 permissible pairs of values of the real variables r, s, and a complex surface has ∞^4 points, corresponding to the ∞^4 sets of values of the four real parameters involved in u, v.

Equation (25), when $U(u)$, $V(v)$ represent triples of analytic functions of the complex variables u and v respectively, represents a complex translation surface. Inasmuch as u_0 and v_0 in (24) are complex parameters, each depending on two real parameters, each family of generators on the surface consists of ∞^2 complex curves.

In order that the surface contain ∞^2 real points, we assume that $V = \overline{U}$, that is, that the functions which V_1, V_2, V_3 are of v are conjugate-complex to the functions which U_1, U_2, U_3 are of u. The surface has, then, the representation

$$x = U(u) + \overline{U}(v). \tag{26}$$

To show that it has ∞^2 real points, we note that, if $t = r + si$ and $\bar{t} = r - si$ are conjugate-complex parameters, corresponding components of $U(t)$ and $\overline{U}(\bar{t})$ are conjugate-complex and the components of $U(t) + \overline{U}(\bar{t})$ are therefore real. Hence every point of the surface whose coordinates (u, v) are of the form $u = t$, $v = \bar{t}$ is a real point. But t depends on the two real parameters r, s. Thus the surface contains the ∞^2 real points represented by the symbolic equation

$$x = U(t) + \overline{U}(\bar{t}). \tag{27}$$

The two generators which pass through the real point $(t_0, \bar{t}_0)$ of the surface (26) have the representations

$$y = U(u) + \overline{U}(\bar{t}_0), \qquad z = U(t_0) + \overline{U}(v). \tag{28}$$

Equivalent representations, obtained by the changes of parameters $u = w$ and $v = \bar{w}$, are

$$y = U(w) + \overline{U}(\bar{t}_0), \qquad z = U(t_0) + \overline{U}(\bar{w}).$$

Since the corresponding components of $U(w) + \overline{U}(\bar{t}_0)$ and $U(t_0) + \overline{U}(\bar{w})$ are conjugate-complex, these representations pair the points of the two generators so that the corresponding coordinates of the points of each pair are conjugate-complex. To express this fact, we say that the points of each pair are conjugate-complex points and that the two generators are conjugate-complex curves.

When t_0 and $\bar{t}_0$ take on all permissible values, equations (28) represent all the generators. Hence the generators of the surface (26) are conjugate-complex in pairs, and each two conjugate-complex generators intersect in a real point (27) of the surface.

From the agreements made at the beginning of § 46, it follows that Theorems 2 and 3 hold for the translation surface (26) as well as for translation surfaces with real generators.

Relationship to harmonic functions. In § 47, we noted that if $F(t)$ is an analytic function of the complex variable $t = r + si$ and we write

$$F(t) = f_1(r, s) + if_2(r, s),$$

then the real functions f_1, f_2 of the real variables r, s are conjugate functions:

$$\frac{\partial f_1}{\partial r} = \frac{\partial f_2}{\partial s}, \quad \frac{\partial f_1}{\partial s} = -\frac{\partial f_2}{\partial r}. \tag{29}$$

Differentiating the first of these equations with respect to r and the second with respect to s and adding the resulting equations, we find that $f_1(r, s)$ satisfies the differential equation of Laplace:

$$\frac{\partial^2 f}{\partial r^2} + \frac{\partial^2 f}{\partial s^2} = 0. \tag{30}$$

Conversely, if $f_1(r, s)$ is an analytic solution of this equation, the equations (29) determine $f_2(r, s)$ to within an additive constant. The functions f_1, f_2 are, then, conjugate functions, and $F = f_1 + if_2$ is an analytic function of the complex variable $r + si$.

An analytic function $f(r, s)$ which satisfies equation (30) is known as a *harmonic function.* Hence we may say: *A real function of the real variables r, s is the real part of an analytic function of the complex variable $r + si$ if and only if it is a harmonic function.*

We return now to equation (27). Inasmuch as

$$U(t) = R(r, s) + iS(r, s), \qquad \overline{U}(\bar{t}) = R(r, s) - iS(r, s),$$

this equation is equivalent to

$$x = 2\,R(r, s). \tag{31}$$

Thus, *we have a representation of the real points* (27) *of the surface* (26) *in terms of real functions of real variables.* Furthermore, it is readily verified that, when r and s are allowed to take on imaginary as well as real values, this representation yields all the points of the surface (26) and no other points.

The real functions $R_1(r, s)$, $R_2(r, s)$, $R_3(r, s)$ are the real parts of the complex functions $U_1(t)$, $U_2(t)$, $U_3(t)$. Hence we conclude:

THEOREM 4. *A necessary and sufficient condition that a surface be a translation surface with imaginary generators is that it admit a representation of the form $x = x(u, v)$, where $x_1(u, v)$, $x_2(u, v)$, $x_3(u, v)$ are harmonic functions.*

EXERCISES

1. Prove that an elliptic paraboloid is a translation surface and identify the generators.

2. The same for a hyperbolic paraboloid.

3. The right helicoid, $x_1 = r\cos\theta$, $x_2 = r\sin\theta$, $x_3 = a\theta$, is in infinitely many ways a translation surface with real generators and in infinitely many ways a translation surface with imaginary generators. In particular, if C is an arbitrary circular helix on the helicoid and K is the circular cylinder on which

C lies, the locus of the mid-points of the chords joining real distinct points of C is the portion of the helicoid inside K, whereas the locus of the mid-points of the chords joining the pairs of conjugate-imaginary points of C is the portion of the helicoid outside K. The generators are real on the former portion and imaginary on the latter.

4. The real curve C: $y = 2\,U(u)$ divides the real part of the complex surface $x = U(u) + U(v)$ into two regions, the region of points for which u, v are real but unequal and the region of points for which u, v are conjugate-imaginary. The points of the first region are the mid-points of the chords joining real distinct points of C, and those of the second are the mid-points of the chords joining conjugate-imaginary points of C. The curve C is an asymptotic line on the surface.

5. Show that a translation surface whose generators are plane curves lying in perpendicular planes may be represented by equations of the form (25) where $V_1 \equiv 0$ and $U_2 \equiv 0$. Show that the asymptotic lines can be found by a quadrature.

6. Show that, if the generators of a translation surface form an orthogonal system, the surface is a cylinder and the generators are the rulings and their orthogonal trajectories.

72. Minimal surfaces. A surface whose mean curvature is zero is, by definition, a minimal surface. If the total curvature is also zero, the surface is a plane, since the vanishing of both the sum and the product of the principal normal curvatures implies that these curvatures are both zero. Hence a minimal surface is an ordinary surface or a plane, never a developable. Excluding planes, we may then restrict ourselves to ordinary surfaces.

THEOREM 1. *A necessary and sufficient condition that a surface be minimal is that the isotropic curves on it form a conjugate system.*

In § 47, we noted that there are always two conjugate-imaginary families of isotropic curves on a surface. Since an isotropic curve is characterized by the property that its differential of arc is zero, the parametric curves on the surface are the isotropic curves if and only if $E = 0$, $G = 0$, $F \neq 0$. But then the condition, $Eg - 2\,Ff + Ge = 0$, that the surface be minimal reduces to the condition, $f = 0$, that the parametric curves form a conjugate system. Hence the theorem is proved.

THEOREM 2. *The minimal surfaces are the translation surfaces with isotropic curves as generators.*

Since the generators of a translation surface always form a conjugate system, a translation surface of the type described is, according to Theorem 1, a minimal surface.

In proving the converse, we assume that the isotropic curves of the given minimal surface are the parametric curves. Then $E = 0$, $G = 0$, and $f = 0$, and it follows from equations (2) and (5) of Chapter VI that $x_{uv} = 0$. Integrating this symbolic differential equation, we find that the parametric representation of the surface must be of the form $x = U(u) + V(v)$. Hence the surface is a translation surface with the isotropic curves on it as generators.

Inasmuch as isotropic curves are always imaginary, we shall write the parametric representation in the form $x = U(u) + \overline{U}(v)$, in keeping with the notation of § 71. We may, then, state our result as follows.

THEOREM 3. *A surface is a minimal surface if and only if it admits a parametric representation of the form*

$$x = U(u) + \overline{U}(v), \tag{32}$$

where $y = U(u)$, $z = \overline{U}(v)$ *represent conjugate-imaginary isotropic curves.*

The equivalent conditions in terms of real functions of the variables r, s, where $u = r + si$, $v = r - si$, are readily found. We know that a representation of the form (32) is equivalent to one of the form $x = 2\,R(r, s)$, where the functions R are harmonic. The requirement that the generators of (32) be isotropic curves is equivalent to the demand that the linear element be of the form $ds^2 = \lambda du dv$ and hence, inasmuch as $dudv = dr^2 + ds^2$, to the demand that r, s be isometric parameters. The desired conditions read, then, as follows.

THEOREM 4. *A surface is minimal if and only if it admits a parametric representation of the form* $x = x(u, v)$, *where* u *and* v *are isometric parameters and* x_1, x_2, x_3 *are harmonic functions of these parameters.*

In order to make equations (32) more explicit, we establish next a characteristic representation of an isotropic curve.

LEMMA. *The isotropic curves, not straight lines, are the curves represented by the symbolic equation*

$$x = \int \phi(u) M(u) du, \tag{33}$$

where $\phi(u)$ is an arbitrary analytic function of the complex parameter u and $M(u)$ is symbolic for the triple of functions

$$(34) \quad M_1(u) = \frac{1 - u^2}{2}, \quad M_2(u) = i\frac{1 + u^2}{2}, \quad M_3(u) = u.$$

The complex analytic curve $x = x(t)$ is an isotropic curve when

$$(35) \qquad x_1'^2 + x_2'^2 + x_3'^2 = 0.$$

Without loss of generality we may assume that x_3' is never zero, that is, that the tangent to the curve is never parallel to the (x_1, x_2)-plane. It follows, then, from (35), that $x_1' - ix_2'$ never vanishes. Hence we may rewrite (35) in the form

$$\frac{x_1' + ix_2'}{-x_3'} = \frac{x_3'}{x_1' - ix_2'}.$$

Denoting by u the common value of these two quotients, we readily find that

$$(36) \qquad \frac{x_1'}{1 - u^2} = \frac{x_2'}{i(1 + u^2)} = \frac{x_3'}{2u}.$$

If u were a constant, this continued equality would tell us that the curve was a straight line. Hence u depends on t and may be introduced as the parameter in place of t. Inasmuch as the triples dx/dt and dx/du are proportional, the continued equality (36) still holds when we interpret x' as dx/du instead of dx/dt. Denoting the common value of the three quotients by $\phi(u)/2$, we readily deduce equation (33). Thus, the lemma is established.

From Theorem 3 and the lemma, we obtain the following result, due to Enneper.

THEOREM 5. *The minimal surfaces are the surfaces represented by the symbolic equation*

$$(37) \qquad x = \int \phi(u)M(u)du + \int \bar{\phi}(v)\bar{M}(v)dv,$$

where $\phi(u)$ and $\bar{\phi}(v)$ are arbitrary conjugate-imaginary analytic functions, $M(u)$ is the triple of functions (34), *and $\bar{M}(v)$ the triple of conjugate-imaginary functions.*

From (37), we find that $x_u = \phi(u)M(u)$, $x_v = \bar{\phi}(v)\bar{M}(v)$. Hence, since

$$(M\,|\,M) = 0, \quad (M\,|\,\bar{M}) = \tfrac{1}{2}(1 + uv)^2, \quad (\bar{M}\,|\,\bar{M}) = 0,$$

we have

$$(38) \quad E = 0, \quad F = \tfrac{1}{2}(1 + uv)^2\phi(u)\bar{\phi}(v), \quad G = 0, \quad D^2 = -F^2.$$

Taking $D = iF$, we get for the components of ζ:

$$(39) \qquad \zeta_1 = \frac{u + v}{1 + uv}, \quad \zeta_2 = i\frac{v - u}{1 + uv}, \quad \zeta_3 = \frac{uv - 1}{1 + uv}.$$

From $e = (\zeta\,|\,x_{uu})$ and the corresponding formulas for f and g, we have

$$(40) \qquad e = -\phi(u), \qquad f = 0, \qquad g = -\bar{\phi}(v).$$

Thus we find that

$$(41) \qquad K = -\frac{1}{\rho^2}, \qquad \rho = \tfrac{1}{2}(1 + uv)^2\sqrt{\phi(u)\bar{\phi}(v)},$$

where $1/\rho$ is the absolute value of each principal normal curvature.

Lines of curvature. The differential equation of the lines of curvature is found to be

$$\phi(u)du^2 - \bar{\phi}(v)dv^2 = 0.$$

Since the variables here are separated, *the lines of curvature can be found by a quadrature.* Their finite equations are $u_1 =$ const. and $v_1 =$ const., where

$$\sqrt{2}\,du_1 = \sqrt{\phi}\,du + \sqrt{\bar{\phi}}\,dv, \qquad \sqrt{2}\,dv_1 = i\sqrt{\phi}\,du - i\sqrt{\bar{\phi}}\,dv.$$

The lines of curvature become the parametric curves when u_1, v_1 are introduced as parameters. If we agree, as is natural, to understand by $\sqrt{\phi}$ and $\sqrt{\bar{\phi}}$ conjugate imaginary square roots of ϕ and $\bar{\phi}$, these parameters are evidently real. Inasmuch as

$$du_1^2 + dv_1^2 = 2\sqrt{\phi\bar{\phi}}\,du\,dv,$$

the linear element, $ds^2 = (1 + uv)^2\,\phi\,\bar{\phi}\,du\,dv$, becomes, in terms of them,

$$ds^2 = \rho(du_1^2 + dv_1^2). \tag{42}$$

Hence we have arrived at the following result.

THEOREM 6. *The lines of curvature on a minimal surface form an isometric system.*

Asymptotic lines. We already know that the asymptotic lines form an orthogonal system. Since their differential equation is

$$\phi(u)du^2 + \bar{\phi}(v)dv^2 = 0, \tag{43}$$

they can be found by the same quadrature as the lines of curvature.

As a matter of fact, it is simple to find parameters for the asymptotic lines in terms of the parameters u_1, v_1 for the lines of curvature. In terms of du_1, dv_1, equation (43) becomes $du_1^2 - dv_1^2 = 0$. Hence we may take as the desired parameters u_2, v_2, where $\sqrt{2}\,u_2 = u_1 - v_1$ and $\sqrt{2}\,v_2 = u_1 + v_1$. In terms of these parameters, the linear element (42) takes the form

$$ds^2 = \rho(du_2^2 + dv_2^2). \tag{44}$$

THEOREM 7. *The asymptotic lines on a minimal surface form an isometric system.*

The theorem is an immediate consequence of (44).

Spherical representation. In § 58 we learned that a minimal surface is mapped conformally on its spherical representation and that, if ds^2 and $d\sigma^2$ are respectively the linear elements of the surface and spherical representation,

$$d\sigma^2 = -\,Kds^2 = \frac{1}{\rho^2}\,ds^2. \tag{45}$$

From (38) and (41) it follows, then, that

$$\mathcal{E} = 0, \qquad \mathcal{F} = \frac{2}{(1 + uv)^2}, \qquad \mathcal{G} = 0.$$

Since $\mathcal{E} = 0$ and $\mathcal{G} = 0$, the isotropic curves on the spherical representation are parametric. But, it is well known that the isotropic curves on a sphere are the rulings on the sphere; see also Ex. 1. Hence, *the spherical representation of the isotropic curves on a minimal surface consists of the isotropic lines on the sphere.*

Incidentally, we have proved that equations (39) constitute a parametric representation of the unit sphere referred to its rulings as the parametric curves.

Inasmuch as a conformal map preserves isometric systems, it follows from Theorems 6 and 7 that *the lines of curvature and the asymptotic lines on a minimal surface are both represented by isometric systems of curves on the sphere.* The forms of the linear element of the sphere corresponding to (42) and (44) are respectively, according to (45),

$$d\sigma^2 = \frac{1}{\rho}(du_1^2 + dv_1^2), \qquad d\sigma^2 = \frac{1}{\rho}(du_2^2 + dv_2^2).$$

Associate minimal surfaces. The family of minimal surfaces

$$z = e^{\alpha i}\int \phi(u)M(u)du + e^{-\alpha i}\int \overline{\phi}(v)\overline{M}(v)dv, \tag{46}$$

where α is a real parameter, is known as a *family of associate minimal surfaces.* Since $e^{\alpha i} = \cos\alpha + i\sin\alpha$, we may, without loss of generality, restrict α to lie in the interval $0 \leqq \alpha < 2\pi$.

The surface $\alpha = 0$ is the surface (37). Since $e^{\pi i} = e^{-\pi i} = -1$, the surface $\alpha = \pi$ is the reflection of this surface in the origin. In fact, each two surfaces of the family corresponding to values of α which differ by π are reflections of one another in the origin.

The fundamental quantities for the general surface (46) of the family are obtainable from those of the particular surface (37) by replacing $\phi(u)$ and $\overline{\phi}(v)$ respectively by $e^{\alpha i}\phi(u)$ and $e^{-\alpha i}\overline{\phi}(v)$. Thereby, e and g are multiplied respectively by $e^{\alpha i}$ and $e^{-\alpha i}$. *All the other quantities remain unchanged.*

This fact has numerous important consequences. Since the components of ζ remain the same, the normals to each two surfaces of the family at corresponding points, that is, points with the same curvilinear coordinates, are parallel. Furthermore, since E, F, G are unchanged, the correspondence between the two surfaces is isometric. Hence we may say:

THEOREM 8. *Each two associate minimal surfaces are mapped isometrically upon one another so that the tangent planes at corresponding points are parallel.*

COROLLARY. *Corresponding directions at corresponding points of the two surfaces make with one another a constant angle. If the surfaces are* $\alpha = \alpha_1$, *and* $\alpha = \alpha_2$, *this angle is* $\alpha_2 - \alpha_1$.

It suffices to prove the corollary for the particular surface (37) and the general surface (46) of the family. At corresponding points of these surfaces,

$$dx = x_u du + x_v dv, \qquad dz = e^{\alpha i} x_u du + e^{-\alpha i} x_v dv,$$

and therefore

$$(dx \,|\, dz) = \frac{(e^{\alpha i} + e^{-\alpha i})}{2} ds^2,$$

where ds^2 is the common linear element of the two surfaces. Hence,

$$\left(\frac{dx}{ds} \middle| \frac{dz}{ds}\right) = \cos \alpha,$$

and the corollary is established.

It follows that corresponding directions at corresponding points of two surfaces of the family which correspond to two values of α differing by $\pi/2$ are always perpendicular to one another. Two such associate minimal surfaces are known as *adjoint minimal surfaces.*

As a typical pair of adjoint surfaces we may take the surface (37) and the surface $\alpha = \pi/2$:

$$y = i\int \phi(u)M(u)du - i\int \bar{\phi}(v)\bar{M}(v)dv. \tag{47}$$

Setting $e^{\alpha i} = \cos \alpha + i \sin \alpha$ and $e^{-\alpha i} = \cos \alpha - i \sin \alpha$ in (46), and collecting the terms in $\cos \alpha$ and $\sin \alpha$ respectively, we readily find that

$$z = x \cos \alpha + y \sin \alpha. \tag{48}$$

The coordinates z of an arbitrary point on the general surface of the family are thus expressed in terms of the coordinates x and y of the corresponding points of the two adjoint surfaces (37) and (47).

It is readily shown that equation (48), when u, v are fixed and α varies, represents an ellipse lying in the plane $(X\, x\, y) = 0$ and having its center in the origin. In other words:

THEOREM 9. *When a minimal surface traces the family of minimal surfaces associate to it, each point on it traces an ellipse.*

The family of surfaces is obviously a continuous family and each two surfaces of it are applicable, by Theorem 8. Hence we may restate the result just obtained in the following way.

THEOREM 10. *A minimal surface admits a continuous deformation whereby it always remains minimal and each point of it traces an ellipse.*

EXERCISES

1. Compute ζ_u and ζ_v from (39). Hence show that the parametric curves on the sphere are isotropic lines and that each isotropic direction at a point of the surface is parallel to the isotropic direction at the image point on the sphere to which it does not correspond.

2. Prove the facts stated in the text concerning equation (48), when u, v are fixed and α varies.

3. The surface of Scherk $x_3 = \log(\cos x_2/\cos x_1)$ is a minimal surface and hence a translation surface with imaginary generators. It is also a translation surface with real generators which lie in two families of perpendicular planes.

4. Show that when $\phi(u) = -a/(2u^2)$, where a is a real constant, the surface (37) is a catenoid.

Suggestion. After evaluating the integrals involved, set

$$u = e^{-\bar{t}} = e^{-r}(\cos s + i \sin s), \quad v = e^{-t} = e^{-r}(\cos s - i \sin s).$$

5. Prove that when $\phi(u) = -a/(2u^2)$, where a is a real constant, the family of associate minimal surfaces (47) can be represented by the equations

$$\begin{aligned} x_1 &= a(\cos\alpha \cosh r \cos s + \sin\alpha \sinh r \sin s), \\ x_2 &= a(\cos\alpha \cosh r \sin s - \sin\alpha \sinh r \cos s), \\ x_3 &= a(r \cos\alpha + s \sin\alpha). \end{aligned}$$

According to Ex. 4, the surface $\alpha = 0$ is a catenoid. Show that the surface $\alpha = \pi/2$ is a right helicoid, thus proving that the catenoid and right helicoid are adjoint surfaces.

6. Show that the surfaces of Ex. 5, where a and α are arbitrary constants, are the ∞^2 helicoidal minimal surfaces. See § 69, Ex. 11.

INDEX

A CATALOG OF SELECTED

DOVER BOOKS

IN SCIENCE AND MATHEMATICS

Astronomy

BURNHAM'S CELESTIAL HANDBOOK, Robert Burnham, Jr. Thorough guide to the stars beyond our solar system. Exhaustive treatment. Alphabetical by constellation: Andromeda to Cetus in Vol. 1; Chamaeleon to Orion in Vol. 2; and Pavo to Vulpecula in Vol. 3. Hundreds of illustrations. Index in Vol. 3. 2,000pp. 6⅛ x 9¼.
Vol. I: 0-486-23567-X
Vol. II: 0-486-23568-8
Vol. III: 0-486-23673-0

EXPLORING THE MOON THROUGH BINOCULARS AND SMALL TELESCOPES, Ernest H. Cherrington, Jr. Informative, profusely illustrated guide to locating and identifying craters, rills, seas, mountains, other lunar features. Newly revised and updated with special section of new photos. Over 100 photos and diagrams. 240pp. 8¼ x 11. 0-486-24491-1

THE EXTRATERRESTRIAL LIFE DEBATE, 1750–1900, Michael J. Crowe. First detailed, scholarly study in English of the many ideas that developed from 1750 to 1900 regarding the existence of intelligent extraterrestrial life. Examines ideas of Kant, Herschel, Voltaire, Percival Lowell, many other scientists and thinkers. 16 illustrations. 704pp. 5⅜ x 8½. 0-486-40675-X

THEORIES OF THE WORLD FROM ANTIQUITY TO THE COPERNICAN REVOLUTION, Michael J. Crowe. Newly revised edition of an accessible, enlightening book recreates the change from an earth-centered to a sun-centered conception of the solar system. 242pp. 5⅜ x 8½. 0-486-41444-2

A HISTORY OF ASTRONOMY, A. Pannekoek. Well-balanced, carefully reasoned study covers such topics as Ptolemaic theory, work of Copernicus, Kepler, Newton, Eddington's work on stars, much more. Illustrated. References. 521pp. 5⅜ x 8½.
0-486-65994-1

A COMPLETE MANUAL OF AMATEUR ASTRONOMY: TOOLS AND TECHNIQUES FOR ASTRONOMICAL OBSERVATIONS, P. Clay Sherrod with Thomas L. Koed. Concise, highly readable book discusses: selecting, setting up and maintaining a telescope; amateur studies of the sun; lunar topography and occultations; observations of Mars, Jupiter, Saturn, the minor planets and the stars; an introduction to photoelectric photometry; more. 1981 ed. 124 figures. 25 halftones. 37 tables. 335pp. 6½ x 9¼. 0-486-40675-X

AMATEUR ASTRONOMER'S HANDBOOK, J. B. Sidgwick. Timeless, comprehensive coverage of telescopes, mirrors, lenses, mountings, telescope drives, micrometers, spectroscopes, more. 189 illustrations. 576pp. 5⅜ x 8¼. (Available in U.S. only.)
0-486-24034-7

STARS AND RELATIVITY, Ya. B. Zel'dovich and I. D. Novikov. Vol. 1 of *Relativistic Astrophysics* by famed Russian scientists. General relativity, properties of matter under astrophysical conditions, stars, and stellar systems. Deep physical insights, clear presentation. 1971 edition. References. 544pp. 5⅜ x 8¼. 0-486-69424-0

Chemistry

THE SCEPTICAL CHYMIST: THE CLASSIC 1661 TEXT, Robert Boyle. Boyle defines the term "element," asserting that all natural phenomena can be explained by the motion and organization of primary particles. 1911 ed. viii+232pp. 5⅜ x 8½.
0-486-42825-7

RADIOACTIVE SUBSTANCES, Marie Curie. Here is the celebrated scientist's doctoral thesis, the prelude to her receipt of the 1903 Nobel Prize. Curie discusses establishing atomic character of radioactivity found in compounds of uranium and thorium; extraction from pitchblende of polonium and radium; isolation of pure radium chloride; determination of atomic weight of radium; plus electric, photographic, luminous, heat, color effects of radioactivity. ii+94pp. 5⅜ x 8½. 0-486-42550-9

CHEMICAL MAGIC, Leonard A. Ford. Second Edition, Revised by E. Winston Grundmeier. Over 100 unusual stunts demonstrating cold fire, dust explosions, much more. Text explains scientific principles and stresses safety precautions. 128pp. 5⅜ x 8½. 0-486-67628-5

THE DEVELOPMENT OF MODERN CHEMISTRY, Aaron J. Ihde. Authoritative history of chemistry from ancient Greek theory to 20th-century innovation. Covers major chemists and their discoveries. 209 illustrations. 14 tables. Bibliographies. Indices. Appendices. 851pp. 5⅜ x 8½. 0-486-64235-6

CATALYSIS IN CHEMISTRY AND ENZYMOLOGY, William P. Jencks. Exceptionally clear coverage of mechanisms for catalysis, forces in aqueous solution, carbonyl- and acyl-group reactions, practical kinetics, more. 864pp. 5⅜ x 8½.
0-486-65460-5

ELEMENTS OF CHEMISTRY, Antoine Lavoisier. Monumental classic by founder of modern chemistry in remarkable reprint of rare 1790 Kerr translation. A must for every student of chemistry or the history of science. 539pp. 5⅜ x 8½. 0-486-64624-6

THE HISTORICAL BACKGROUND OF CHEMISTRY, Henry M. Leicester. Evolution of ideas, not individual biography. Concentrates on formulation of a coherent set of chemical laws. 260pp. 5⅜ x 8½. 0-486-61053-5

A SHORT HISTORY OF CHEMISTRY, J. R. Partington. Classic exposition explores origins of chemistry, alchemy, early medical chemistry, nature of atmosphere, theory of valency, laws and structure of atomic theory, much more. 428pp. 5⅜ x 8½. (Available in U.S. only.) 0-486-65977-1

GENERAL CHEMISTRY, Linus Pauling. Revised 3rd edition of classic first-year text by Nobel laureate. Atomic and molecular structure, quantum mechanics, statistical mechanics, thermodynamics correlated with descriptive chemistry. Problems. 992pp. 5⅜ x 8½. 0-486-65622-5

FROM ALCHEMY TO CHEMISTRY, John Read. Broad, humanistic treatment focuses on great figures of chemistry and ideas that revolutionized the science. 50 illustrations. 240pp. 5⅜ x 8½. 0-486-28690-8

Engineering

DE RE METALLICA, Georgius Agricola. The famous Hoover translation of greatest treatise on technological chemistry, engineering, geology, mining of early modern times (1556). All 289 original woodcuts. 638pp. 6¾ x 11. 0-486-60006-8

FUNDAMENTALS OF ASTRODYNAMICS, Roger Bate et al. Modern approach developed by U.S. Air Force Academy. Designed as a first course. Problems, exercises. Numerous illustrations. 455pp. 5⅜ x 8½. 0-486-60061-0

DYNAMICS OF FLUIDS IN POROUS MEDIA, Jacob Bear. For advanced students of ground water hydrology, soil mechanics and physics, drainage and irrigation engineering and more. 335 illustrations. Exercises, with answers. 784pp. 6⅛ x 9¼. 0-486-65675-6

THEORY OF VISCOELASTICITY (Second Edition), Richard M. Christensen. Complete consistent description of the linear theory of the viscoelastic behavior of materials. Problem-solving techniques discussed. 1982 edition. 29 figures. xiv+364pp. 6⅛ x 9¼. 0-486-42880-X

MECHANICS, J. P. Den Hartog. A classic introductory text or refresher. Hundreds of applications and design problems illuminate fundamentals of trusses, loaded beams and cables, etc. 334 answered problems. 462pp. 5⅜ x 8½. 0-486-60754-2

MECHANICAL VIBRATIONS, J. P. Den Hartog. Classic textbook offers lucid explanations and illustrative models, applying theories of vibrations to a variety of practical industrial engineering problems. Numerous figures. 233 problems, solutions. Appendix. Index. Preface. 436pp. 5⅜ x 8½. 0-486-64785-4

STRENGTH OF MATERIALS, J. P. Den Hartog. Full, clear treatment of basic material (tension, torsion, bending, etc.) plus advanced material on engineering methods, applications. 350 answered problems. 323pp. 5⅜ x 8½. 0-486-60755-0

A HISTORY OF MECHANICS, René Dugas. Monumental study of mechanical principles from antiquity to quantum mechanics. Contributions of ancient Greeks, Galileo, Leonardo, Kepler, Lagrange, many others. 671pp. 5⅜ x 8½. 0-486-65632-2

STABILITY THEORY AND ITS APPLICATIONS TO STRUCTURAL MECHANICS, Clive L. Dym. Self-contained text focuses on Koiter postbuckling analyses, with mathematical notions of stability of motion. Basing minimum energy principles for static stability upon dynamic concepts of stability of motion, it develops asymptotic buckling and postbuckling analyses from potential energy considerations, with applications to columns, plates, and arches. 1974 ed. 208pp. 5⅜ x 8½. 0-486-42541-X

METAL FATIGUE, N. E. Frost, K. J. Marsh, and L. P. Pook. Definitive, clearly written, and well-illustrated volume addresses all aspects of the subject, from the historical development of understanding metal fatigue to vital concepts of the cyclic stress that causes a crack to grow. Includes 7 appendixes. 544pp. 5⅜ x 8½. 0-486-40927-9

ROCKETS, Robert Goddard. Two of the most significant publications in the history of rocketry and jet propulsion: "A Method of Reaching Extreme Altitudes" (1919) and "Liquid Propellant Rocket Development" (1936). 128pp. 5⅜ x 8½. 0-486-42537-1

STATISTICAL MECHANICS: PRINCIPLES AND APPLICATIONS, Terrell L. Hill. Standard text covers fundamentals of statistical mechanics, applications to fluctuation theory, imperfect gases, distribution functions, more. 448pp. 5⅜ x 8½. 0-486-65390-0

ENGINEERING AND TECHNOLOGY 1650–1750: ILLUSTRATIONS AND TEXTS FROM ORIGINAL SOURCES, Martin Jensen. Highly readable text with more than 200 contemporary drawings and detailed engravings of engineering projects dealing with surveying, leveling, materials, hand tools, lifting equipment, transport and erection, piling, bailing, water supply, hydraulic engineering, and more. Among the specific projects outlined-transporting a 50-ton stone to the Louvre, erecting an obelisk, building timber locks, and dredging canals. 207pp. 8⅜ x 11¼. 0-486-42232-1

THE VARIATIONAL PRINCIPLES OF MECHANICS, Cornelius Lanczos. Graduate level coverage of calculus of variations, equations of motion, relativistic mechanics, more. First inexpensive paperbound edition of classic treatise. Index. Bibliography. 418pp. 5⅜ x 8½. 0-486-65067-7

PROTECTION OF ELECTRONIC CIRCUITS FROM OVERVOLTAGES, Ronald B. Standler. Five-part treatment presents practical rules and strategies for circuits designed to protect electronic systems from damage by transient overvoltages. 1989 ed. xxiv+434pp. 6⅛ x 9¼. 0-486-42552-5

ROTARY WING AERODYNAMICS, W. Z. Stepniewski. Clear, concise text covers aerodynamic phenomena of the rotor and offers guidelines for helicopter performance evaluation. Originally prepared for NASA. 537 figures. 640pp. 6⅛ x 9¼. 0-486-64647-5

INTRODUCTION TO SPACE DYNAMICS, William Tyrrell Thomson. Comprehensive, classic introduction to space-flight engineering for advanced undergraduate and graduate students. Includes vector algebra, kinematics, transformation of coordinates. Bibliography. Index. 352pp. 5⅜ x 8½. 0-486-65113-4

HISTORY OF STRENGTH OF MATERIALS, Stephen P. Timoshenko. Excellent historical survey of the strength of materials with many references to the theories of elasticity and structure. 245 figures. 452pp. 5⅜ x 8½. 0-486-61187-6

ANALYTICAL FRACTURE MECHANICS, David J. Unger. Self-contained text supplements standard fracture mechanics texts by focusing on analytical methods for determining crack-tip stress and strain fields. 336pp. 6⅛ x 9¼. 0-486-41737-9

STATISTICAL MECHANICS OF ELASTICITY, J. H. Weiner. Advanced, self-contained treatment illustrates general principles and elastic behavior of solids. Part 1, based on classical mechanics, studies thermoelastic behavior of crystalline and polymeric solids. Part 2, based on quantum mechanics, focuses on interatomic force laws, behavior of solids, and thermally activated processes. For students of physics and chemistry and for polymer physicists. 1983 ed. 96 figures. 496pp. 5⅜ x 8½. 0-486-42260-7

CALCULUS OF VARIATIONS WITH APPLICATIONS, George M. Ewing. Applications-oriented introduction to variational theory develops insight and promotes understanding of specialized books, research papers. Suitable for advanced undergraduate/graduate students as primary, supplementary text. 352pp. 5⅜ x 8½. 0-486-64856-7

COMPLEX VARIABLES, Francis J. Flanigan. Unusual approach, delaying complex algebra till harmonic functions have been analyzed from real variable viewpoint. Includes problems with answers. 364pp. 5⅜ x 8½. 0-486-61388-7

AN INTRODUCTION TO THE CALCULUS OF VARIATIONS, Charles Fox. Graduate-level text covers variations of an integral, isoperimetrical problems, least action, special relativity, approximations, more. References. 279pp. 5⅜ x 8½. 0-486-65499-0

COUNTEREXAMPLES IN ANALYSIS, Bernard R. Gelbaum and John M. H. Olmsted. These counterexamples deal mostly with the part of analysis known as "real variables." The first half covers the real number system, and the second half encompasses higher dimensions. 1962 edition. xxiv+198pp. 5⅜ x 8½. 0-486-42875-3

CATASTROPHE THEORY FOR SCIENTISTS AND ENGINEERS, Robert Gilmore. Advanced-level treatment describes mathematics of theory grounded in the work of Poincaré, R. Thom, other mathematicians. Also important applications to problems in mathematics, physics, chemistry and engineering. 1981 edition. References. 28 tables. 397 black-and-white illustrations. xvii + 666pp. 6⅛ x 9¼. 0-486-67539-4

INTRODUCTION TO DIFFERENCE EQUATIONS, Samuel Goldberg. Exceptionally clear exposition of important discipline with applications to sociology, psychology, economics. Many illustrative examples; over 250 problems. 260pp. 5⅜ x 8½. 0-486-65084-7

NUMERICAL METHODS FOR SCIENTISTS AND ENGINEERS, Richard Hamming. Classic text stresses frequency approach in coverage of algorithms, polynomial approximation, Fourier approximation, exponential approximation, other topics. Revised and enlarged 2nd edition. 721pp. 5⅜ x 8½. 0-486-65241-6

INTRODUCTION TO NUMERICAL ANALYSIS (2nd Edition), F. B. Hildebrand. Classic, fundamental treatment covers computation, approximation, interpolation, numerical differentiation and integration, other topics. 150 new problems. 669pp. 5⅜ x 8½. 0-486-65363-3

THREE PEARLS OF NUMBER THEORY, A. Y. Khinchin. Three compelling puzzles require proof of a basic law governing the world of numbers. Challenges concern van der Waerden's theorem, the Landau-Schnirelmann hypothesis and Mann's theorem, and a solution to Waring's problem. Solutions included. 64pp. 5⅜ x 8½. 0-486-40026-3

THE PHILOSOPHY OF MATHEMATICS: AN INTRODUCTORY ESSAY, Stephan Körner. Surveys the views of Plato, Aristotle, Leibniz & Kant concerning propositions and theories of applied and pure mathematics. Introduction. Two appendices. Index. 198pp. 5⅜ x 8½. 0-486-25048-2

Math–Decision Theory, Statistics, Probability

ELEMENTARY DECISION THEORY, Herman Chernoff and Lincoln E. Moses. Clear introduction to statistics and statistical theory covers data processing, probability and random variables, testing hypotheses, much more. Exercises. 364pp. 5⅜ x 8½. 0-486-65218-1

STATISTICS MANUAL, Edwin L. Crow et al. Comprehensive, practical collection of classical and modern methods prepared by U.S. Naval Ordnance Test Station. Stress on use. Basics of statistics assumed. 288pp. 5⅜ x 8½. 0-486-60599-X

SOME THEORY OF SAMPLING, William Edwards Deming. Analysis of the problems, theory and design of sampling techniques for social scientists, industrial managers and others who find statistics important at work. 61 tables. 90 figures. xvii +602pp. 5⅜ x 8½. 0-486-64684-X

LINEAR PROGRAMMING AND ECONOMIC ANALYSIS, Robert Dorfman, Paul A. Samuelson and Robert M. Solow. First comprehensive treatment of linear programming in standard economic analysis. Game theory, modern welfare economics, Leontief input-output, more. 525pp. 5⅜ x 8½. 0-486-65491-5

PROBABILITY: AN INTRODUCTION, Samuel Goldberg. Excellent basic text covers set theory, probability theory for finite sample spaces, binomial theorem, much more. 360 problems. Bibliographies. 322pp. 5⅜ x 8½. 0-486-65252-1

GAMES AND DECISIONS: INTRODUCTION AND CRITICAL SURVEY, R. Duncan Luce and Howard Raiffa. Superb nontechnical introduction to game theory, primarily applied to social sciences. Utility theory, zero-sum games, n-person games, decision-making, much more. Bibliography. 509pp. 5⅜ x 8½. 0-486-65943-7

INTRODUCTION TO THE THEORY OF GAMES, J. C. C. McKinsey. This comprehensive overview of the mathematical theory of games illustrates applications to situations involving conflicts of interest, including economic, social, political, and military contexts. Appropriate for advanced undergraduate and graduate courses; advanced calculus a prerequisite. 1952 ed. x+372pp. 5⅜ x 8½. 0-486-42811-7

FIFTY CHALLENGING PROBLEMS IN PROBABILITY WITH SOLUTIONS, Frederick Mosteller. Remarkable puzzlers, graded in difficulty, illustrate elementary and advanced aspects of probability. Detailed solutions. 88pp. 5⅜ x 8½. 65355-2

PROBABILITY THEORY: A CONCISE COURSE, Y. A. Rozanov. Highly readable, self-contained introduction covers combination of events, dependent events, Bernoulli trials, etc. 148pp. 5⅜ x 8¼. 0-486-63544-9

STATISTICAL METHOD FROM THE VIEWPOINT OF QUALITY CONTROL, Walter A. Shewhart. Important text explains regulation of variables, uses of statistical control to achieve quality control in industry, agriculture, other areas. 192pp. 5⅜ x 8½. 0-486-65232-7

History of Math

THE WORKS OF ARCHIMEDES, Archimedes (T. L. Heath, ed.). Topics include the famous problems of the ratio of the areas of a cylinder and an inscribed sphere; the measurement of a circle; the properties of conoids, spheroids, and spirals; and the quadrature of the parabola. Informative introduction. clxxxvi+326pp. 5⅜ x 8½. 0-486-42084-1

A SHORT ACCOUNT OF THE HISTORY OF MATHEMATICS, W. W. Rouse Ball. One of clearest, most authoritative surveys from the Egyptians and Phoenicians through 19th-century figures such as Grassman, Galois, Riemann. Fourth edition. 522pp. 5⅜ x 8½. 0-486-20630-0

THE HISTORY OF THE CALCULUS AND ITS CONCEPTUAL DEVELOPMENT, Carl B. Boyer. Origins in antiquity, medieval contributions, work of Newton, Leibniz, rigorous formulation. Treatment is verbal. 346pp. 5⅜ x 8½. 0-486-60509-4

THE HISTORICAL ROOTS OF ELEMENTARY MATHEMATICS, Lucas N. H. Bunt, Phillip S. Jones, and Jack D. Bedient. Fundamental underpinnings of modern arithmetic, algebra, geometry and number systems derived from ancient civilizations. 320pp. 5⅜ x 8½. 0-486-25563-8

A HISTORY OF MATHEMATICAL NOTATIONS, Florian Cajori. This classic study notes the first appearance of a mathematical symbol and its origin, the competition it encountered, its spread among writers in different countries, its rise to popularity, its eventual decline or ultimate survival. Original 1929 two-volume edition presented here in one volume. xxviii+820pp. 5⅜ x 8½. 0-486-67766-4

GAMES, GODS & GAMBLING: A HISTORY OF PROBABILITY AND STATISTICAL IDEAS, F. N. David. Episodes from the lives of Galileo, Fermat, Pascal, and others illustrate this fascinating account of the roots of mathematics. Features thought-provoking references to classics, archaeology, biography, poetry. 1962 edition. 304pp. 5⅜ x 8½. (Available in U.S. only.) 0-486-40023-9

OF MEN AND NUMBERS: THE STORY OF THE GREAT MATHEMATICIANS, Jane Muir. Fascinating accounts of the lives and accomplishments of history's greatest mathematical minds–Pythagoras, Descartes, Euler, Pascal, Cantor, many more. Anecdotal, illuminating. 30 diagrams. Bibliography. 256pp. 5⅜ x 8½. 0-486-28973-7

HISTORY OF MATHEMATICS, David E. Smith. Nontechnical survey from ancient Greece and Orient to late 19th century; evolution of arithmetic, geometry, trigonometry, calculating devices, algebra, the calculus. 362 illustrations. 1,355pp. 5⅜ x 8½. Two-vol. set. Vol. I: 0-486-20429-4 Vol. II: 0-486-20430-8

A CONCISE HISTORY OF MATHEMATICS, Dirk J. Struik. The best brief history of mathematics. Stresses origins and covers every major figure from ancient Near East to 19th century. 41 illustrations. 195pp. 5⅜ x 8½. 0-486-60255-9

Physics

OPTICAL RESONANCE AND TWO-LEVEL ATOMS, L. Allen and J. H. Eberly. Clear, comprehensive introduction to basic principles behind all quantum optical resonance phenomena. 53 illustrations. Preface. Index. 256pp. 5⅜ x 8½. 0-486-65533-4

QUANTUM THEORY, David Bohm. This advanced undergraduate-level text presents the quantum theory in terms of qualitative and imaginative concepts, followed by specific applications worked out in mathematical detail. Preface. Index. 655pp. 5⅜ x 8½. 0-486-65969-0

ATOMIC PHYSICS (8th EDITION), Max Born. Nobel laureate's lucid treatment of kinetic theory of gases, elementary particles, nuclear atom, wave-corpuscles, atomic structure and spectral lines, much more. Over 40 appendices, bibliography. 495pp. 5⅜ x 8½. 0-486-65984-4

A SOPHISTICATE'S PRIMER OF RELATIVITY, P. W. Bridgman. Geared toward readers already acquainted with special relativity, this book transcends the view of theory as a working tool to answer natural questions: What is a frame of reference? What is a "law of nature"? What is the role of the "observer"? Extensive treatment, written in terms accessible to those without a scientific background. 1983 ed. xlviii+172pp. 5⅜ x 8½. 0-486-42549-5

AN INTRODUCTION TO HAMILTONIAN OPTICS, H. A. Buchdahl. Detailed account of the Hamiltonian treatment of aberration theory in geometrical optics. Many classes of optical systems defined in terms of the symmetries they possess. Problems with detailed solutions. 1970 edition. xv + 360pp. 5⅜ x 8½. 0-486-67597-1

PRIMER OF QUANTUM MECHANICS, Marvin Chester. Introductory text examines the classical quantum bead on a track: its state and representations; operator eigenvalues; harmonic oscillator and bound bead in a symmetric force field; and bead in a spherical shell. Other topics include spin, matrices, and the structure of quantum mechanics; the simplest atom; indistinguishable particles; and stationary-state perturbation theory. 1992 ed. xiv+314pp. 6⅛ x 9¼. 0-486-42878-8

LECTURES ON QUANTUM MECHANICS, Paul A. M. Dirac. Four concise, brilliant lectures on mathematical methods in quantum mechanics from Nobel Prize-winning quantum pioneer build on idea of visualizing quantum theory through the use of classical mechanics. 96pp. 5⅜ x 8½. 0-486-41713-1

THIRTY YEARS THAT SHOOK PHYSICS: THE STORY OF QUANTUM THEORY, George Gamow. Lucid, accessible introduction to influential theory of energy and matter. Careful explanations of Dirac's anti-particles, Bohr's model of the atom, much more. 12 plates. Numerous drawings. 240pp. 5⅜ x 8½. 0-486-24895-X

ELECTRONIC STRUCTURE AND THE PROPERTIES OF SOLIDS: THE PHYSICS OF THE CHEMICAL BOND, Walter A. Harrison. Innovative text offers basic understanding of the electronic structure of covalent and ionic solids, simple metals, transition metals and their compounds. Problems. 1980 edition. 582pp. 6⅛ x 9¼. 0-486-66021-4